CEMETERY STORIES

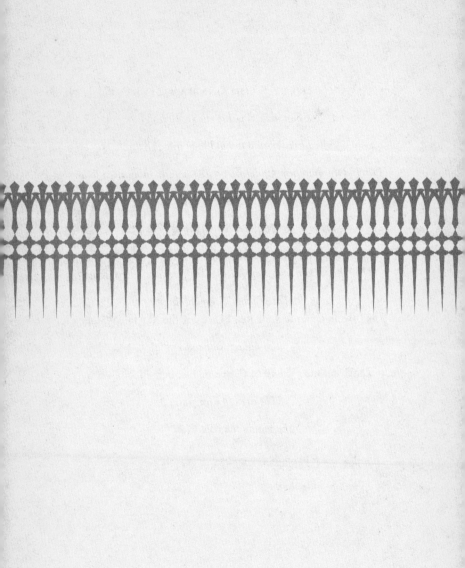

Haunted Graveyards,

Embalming Secrets,

and the Life of a Corpse After Death

CEMETERY
STORIES

Katherine Ramsland

HarperEntertainment
An Imprint of HarperCollinsPublishers

HarperCollins books may be purchased for educational, business,
or sales promotional use. For information please write Special
Markets Department, HarperCollins Publishers Inc., 10 East 53rd
Street, New York, NY 10022.

FIRST EDITION

Designed by Elliott Beard

Printed on acid-free paper.

Library of Congress Cataloging-in-Publication Data
Ramsland, Katherine M., 1953–
    Cemetery stories : haunted graveyards, embalming secrets &
the life of a corpse after death / by Katherine Ramsland.
  —1st ed.
     p. cm.
    Includes bibliographical references.
    ISBN 0-06-018518-X
    1. Funeral rites and ceremonies—United States—Anec-
dotes.   2. Cemeteries—United States—Anecdotes.
3. Dead—Anecdotes. I. Title.
GT3203 .R35 2001
393.1'0973—dc21   2001024499

  02 03 04 05 ❖/RRD 10 9 8 7 6 5 4 3

FOR MY MOTHER,
Who encouraged me to explore,

AND FOR MISS MARY,
Who shares the adventure

Contents

Acknowledgments

For help with stories, introductions, and exposure to cemetery culture, I would like to thank all of the people who are mentioned in the text by name, all of the funeral directors and caretakers who told me stories anonymously, and the following:

Deborah Brown and Jay Burnside were quite generous in helping me to meet people in the funeral industry and to answer my questions.

Others who pointed me in the right direction were Marianne Bergé, George Brown, Doug Clegg, Andrea Del Favero, Michelle Dickens, Mark Diionno, Wanda Hendrix, Barbara Johnston, Donna Johnston, Ruth Osborne, Mary Farrelly, Dorothy Fiedel, Deborah Iida, Michael McCloskey, Mark Nesbitt, Lori Perkins, Mark Spivey, Andrew Saal, Tony Sokol, Richard Kramer, Jeanne Youngson, members of the Grave-L chat list, Laura, Jane, and Tee.

My editor, Josh Behar, kept the fires burning with his tireless enthusiasm and his eager anticipation of the next good story. His vision kept this array of tales organized and his editorial perspective was without equal.

Introduction

Oh, don't you laugh when the hearse goes by
Or you will be the next to die
They wrap you up in bloody sheets
And then they bury you six feet deep
The worms crawl in, the worms crawl out
The worms play pinochle on your snout
There're big green bugs with big green eyes
They go in your nose and out your eyes
And then you mold away.
—child's ditty

When I was five, I often sang this little song with my friends. They always joked about it, but I took it seriously. No way was I ever going to ride off in one of those hearses or be invaded by worms and bugs, and I was certainly not going to mold away.

As I grew older, I learned that this kind of stuff actually does happen, whether I laughed at a hearse or not. So like a lot of people, I tried to avoid funerals, or if I had to go, I avoided the corpse in the casket. Cemeteries to me were just interesting places to walk. I didn't think about what might be happening under the earth.

Then my grandfather died. For the first time I had the opportunity to sit alone with a dead body. I knew that he would soon be shoved into a mausoleum drawer and I wanted some private time with him. He'd been embalmed, although I didn't quite understand what that meant. I just knew that he was lying there stiff and full of something that made his cold skin feel disturbingly hard and waxy. His eyes were closed and he seemed more peaceful than he'd been during his illness, yet still he wore a strained expression. I can't really say that he looked like himself, but I didn't realize then that he'd been calculatedly posed to give me a "memory picture." I also didn't know what would happen to him now and I didn't like to think of him encased in a box in the dark, even if he did have a nice mattress.

After that I avoided funerals again, and even the cemeteries lost some of their allure. But then a close friend died and I had to help make some of the postmortem decisions. Everything happened too quickly and I wished I'd understood more about the process. It was time, I realized, to learn more about the subject of death and burial. I wanted to know what would happen to me and whether I had any choices in the matter. I also wanted to know who these "death-care professionals" were who handled all the procedures right down to the grave.

Then I had an encounter with someone who opened the door.

When I came down for breakfast one morning at a B&B in northern Maryland, it seemed like an ordinary day. Nothing prepared me in that moment for the discussion I was about to have.

The breakfast table had room for six, but only one person sat there with his coffee. I took a seat and introduced myself. My lone companion, Charles Zannino, was a tall, lean man with wavy dark hair, and from the way the innkeeper addressed him when she brought out his pancakes, I sensed he was a regular guest. Greeting me with a warm smile, he mentioned that he had quite a busy day ahead.

"What do you do?" I asked.

"I'm an embalmer," he responded.

Yikes. I wasn't sure what to say, but then I figured, well, here was my chance. I had in front of me someone who could demystify a subject that most people knew nothing about, so I invited him to tell me more.

He was delighted. As I ate my scrambled eggs (sans ketchup), he launched right into a description about how the mummified finger of an older corpse can be hydrated to get fingerprints.

"There are all kinds of reasons why you might find a body in this state," he said. "A homeless person left to the elements, for example, or a murder victim who wasn't found for months."

Despite my past avoidance of death issues, I found myself fascinated. He then initiated a discussion about the popular idea that hair and nails continue to grow after death. They most certainly do not, he insisted.

"Are you sure?" I pressed. I'd once heard about a tornado that had mowed through a cemetery and sucked freshly buried caskets out of the ground. A man's corpse had been found several yards from its resting place, and its hair and nails were six inches longer than when they had buried him. At least that's what I'd been told, and that remained one of the spookiest real-life images I'd ever heard.

When I offered this to Zannino, he laughed. "That's an urban legend," he assured me. I was deflated, but he went on to explain. "The skin on the nails dehydrates and recedes, and the hairline falls back so that it looks like they've grown, but they haven't."

He obviously had a passion for his work, and as he talked that morning, I noted that he was also a keen observer of odd things that happen behind a mortuary's closed doors. "There are a lot of stories," he confided.

That's all he needed to say. Suddenly, I was interested—profoundly interested. I wanted to learn this stuff, but more than that, I wanted to hear some good stories. Where there were bodies, there *had* to be things happening behind the scenes. I decided then and there to seek out some people who could tell me more about the whole process, but also divulge the secrets. In the end, while some of my expectations proved groundless, other things were even creepier than I had thought.

Cemetery stories—the tales that people tell about funerals and graveyards. I figured there were as many as there were people who'd been buried. Yet I soon discovered that getting these stories was no easy task. First, many people in the death-care industry have incredibly busy schedules. Like physicians, they never know when they might be needed, and when a call comes in, they have to go. (One man told me that in twenty-five years of marriage, he had managed to spend only one Christmas with his family.) Numerous interviews were rescheduled and then rescheduled yet again. I came to have a great deal of compassion for how hard funeral directors work and for the kinds of irregular lives they have to lead to serve the public and make a living.

But there was also another issue: While the poet undertaker Thomas Lynch had no trouble telling thought-provoking stories about his trade, others were more guarded. What kind of information was I seeking? Was I going to write about it and make them look like vultures, as had certain authors who had criticized the American way of death? Was I going to disparage the corporate developments in this business or make them all into morbid necrophiles? Would I embarrass them to their clients?

I understood their concerns, yet my intention was to find out what was really going on behind closed doors, from the principled to the unethical. If some funeral director threw a party that included the bodies, I would say so, but I would acknowledge the experiences of the normal professional, too. I wasn't sure what I would discover, yet by the time I had covered the range of things that can happen during the life of a corpse after death, my entire perspective had changed.

People in the death-care business certainly see some interesting things. What I have collected is a wide variety of tales from those who work in, are associated with, love, care for, or have anything to do with funerals and cemeteries. From embalmers to gravediggers to "taphophiles," the surprising things that can happen would wake even the dead.

How did I pick the stories that ended up in these pages? Out of all the cemeteries and funeral homes in the world, and out of the entire history of burying people, what made these stand out? I knew it would be impossible to cover everything, so I allowed the book to form itself via six routes:

1. Serendipity—I was with someone who knew a funeral director who had a good story.

2. Personal interest—such as my quest to find out about Jim Morrison's grave in Paris.

3. Stories sent from friends.

4. Cemeteries that sponsored some event, such as the birthday party for the oldest cemetery in Atlanta.

5. Tales or facts that educated.

As I gathered material from all over the world, this book grew into three natural divisions:

1. Stories told about and by cemetery personnel.

2. Stories about particular cemeteries or graves.

3. Stories about events or people in cemetery culture that are disturbing, scary, or utterly weird.

But where to begin? That was a dilemma. While I made an appointment to talk more at length with Charles Zannino about embalming, I decided to explore my home turf and right away I found a rather surprising cemetery mystery.

As I strolled through the manicured grounds of Princeton Cemetery, I came across an older man tending the trees. His name was George Brown, and he wasn't paid for what he did, but he liked to take care of the place. We walked around together as he talked about what the cemetery meant to him, and it turned out that he was quite the historian.

The main cemetery in the center of Princeton, New Jersey, is quite famous for its distinguished occupants. President Grover Cleveland lies here, as do Aaron Burr and Declaration of Independence signer John Witherspoon. Three Civil War generals each occupy a plot, and one can find scientists, mathematicians, activists, philosophers, and various and sundry writers. This cemetery, only two blocks from the Princeton University campus, has been in operation for over two centuries. One of its unique stories surrounds the tomb of an escaped slave, James Johnson, who died in 1902. Johnson had run away from his master in Maryland and ended up in Princeton. For a while, he survived by selling snacks to students, but eventually someone recognized him and turned him in. A Princeton resident paid for his freedom and he worked hard to repay her. He remained in Princeton until he died, still selling snacks. Rather than leave him in a pauper's grave, several students and alumni took up a collection and erected a monument for him.

George Brown and I continued to walk toward one corner of the cemetery so he could show me a burial spot that few people knew about. He found it interesting that the grave was even here and hinted that there'd been a rather odd event associated with it. The grave wasn't on any cemetery map, and once we arrived I understood why. Right there in front of me were the side-by-side plots of Kitty and José Menendez, victims in one of the most sensational double-murder cases in recent decades. They had been killed in Beverly Hills, California, in their four-million-dollar mansion, yet here they were in Princeton Cemetery. And not on the map. To get the rest of the story, George referred me to someone who preferred to remain anonymous, but who freely talked.

The strange event had occurred back in 1989, I learned, not long after the murders. At the time, all anyone knew was that on the night of August 20, the Menendezes' two sons, Lyle and Erik, had called 911 to report that their parents had been brutally murdered. The investigation made clear that both victims had been shot numerous times with twelve-gauge shotguns, which had been reloaded to shoot again. Kitty had tried crawling away, her body riddled with buckshot, and had been shot ten more times before she finally died. The boys had suggested that this was the work of "the mob."

Apparently in a fit of mourning, they then spent nearly one million dollars over the next four months, buying everything from Rolex watches to cars to a Princeton restaurant. Just before this spree, they had come to Princeton to stage an elaborate memorial service for their departed parents. Part of that involved a meeting with cemetery officials, and this is where the story takes an odd turn. It seems that the Menendez brothers wanted to purchase plots and have Kitty's and José's bodies transported there, since this was where many family members lived. Yet to the surprise of those they contacted, the brothers wanted to purchase not two plots, nor four. They wanted to buy one hundred cemetery plots!

No one could figure out why they wanted so many, and they wouldn't reveal their plans, but in the end they were allowed only two. Princeton was a cemetery in high demand and you couldn't just come in and make a claim on that much space. The deceased Menendezes were duly buried while the search for their killers continued.

Then, in March of 1990, Lyle and Erik were arrested and charged. One of them had let slip to a counselor what they had done. However, at their trial they claimed self-defense and a life of abuse. Nevertheless, their spending spree made it clear that they may have killed their parents for the money they expected to get. That put their cemetery visit into a new light. Had the multiple plots been a guilt offering? We may never know, since both men are currently in prison for life.

Now that's a story . . . and there are plenty more inside.

ONE

Workers of the Dead

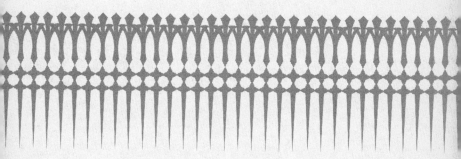

The Burial Detail

Sometimes finding a story means being in the right place at the right time, and one day I came across a rare sight. I was walking in the municipal cemetery in the midwestern town where I grew up. All around me were the typical Victorian-era monuments of various sizes, from granite slabs to statues to marble spires. They each marked underground, grass-covered graves that surrounded a solid stone mausoleum right in the center. The old part, once a potter's field, ran up a sloping hill to my left. The oldest stone dates back to 1825.

What a lot of people don't realize is that in the nineteenth century, townspeople thought nothing of coming to cemeteries like this for picnics because it was a place for exchanging news and gossip while also paying one's respects to the dead. Somehow, solitary family visits to graves eventually eclipsed that tradition, and now many people avoid cemeteries altogether.

Maybe that's due to superstition or the way we've pushed death from our homes and our thoughts. Many religions believe that ghosts haunt cemeteries, and the ones most likely to be hanging around are those that suffered or those seeking revenge. While some ghosts supposedly haunt the place where they died, others might just as easily be near their bodies, emerging confused and disoriented after the body is buried.

I once had that notion. As a child cutting through the very cemetery I was presently in, I was sure that if I stepped on someone's grave, a ghost would hitch a ride home with me. I'd spent many sleepless nights believing that I'd committed a spiritual trespass. I was especially nervous about a pair of double wooden doors built into the side of the hill. Something unspeakable had to be locked inside the earth. Just glancing at those doors still filled me with dread, so I quickly walked away.

I intended to see if anyone I knew had been buried lately. I had a few former friends here already, including the victim of a suicide. I entered by a side gate, and as I rounded a turn on the narrow paved road and came out from behind the mausoleum, I was surprised to see a crew of men dressed in orange uniforms. Then I noticed they were digging a grave!

I immediately recalled the scene in *Hamlet* where the gravedigger engages the Danish prince in a battle of wits. He's seen many an elevated person go down into the dirt, he says, causing Hamlet to ponder "to what base uses we may return." I wondered if any of this crew thought about life and death as they prepared the ground for a body.

Someone operated a backhoe, while a rather formidable man wearing a black blazer stood at the side of the deep hole. As I came closer, I saw two men inside, using shovels to flatten the rich, dark earth. The third worker jumped up and down as if to make it smooth and hard. As he climbed out of the six-foot opening and came over to a building near me, I asked if I could talk to him about what he was doing. He gave me a strange look. Then pointing to the man in black, he said, "Talk to him."

It turned out that the man in black was a corrections officer, and he was there supervising the other men—all of them prisoners. It was then that I noticed seven other men dressed in bright orange vests working around the cemetery. One was whipping weeds from a monument, another mowed the lawn, and others were working on the cement foundation on which a grave marker would eventually be placed. The officer told me that the city contracts with the prison to do various jobs, including digging graves.

"Don't any of them object?" I asked. I mean, you get thrown into prison with no idea that you'd get the cemetery detail. "Some of them must be superstitious."

"They all object," he said with a grim smile, "but they get over that pretty fast."

Just then a good-looking guy in a turquoise T-shirt walked up and introduced himself as Dan Bennett, the cemetery's caretaker. He looked to be approximately thirty, and sported a tiny gold loop pierced through one ear. He'd been in the position for over four years, he told me, and "if you don't think too hard about what you're doing, it's okay." Since he does not have regular gravediggers on staff, he has to rely on prisoners. It took about forty-five minutes to an hour, he estimated, to dig a grave. That was quite a change from the days before machinery.

In the beginning, when someone died, his family or friends did the deed, but eventually gravedigging became a specialized role for someone in a church parish—often the schoolmaster. That started in the 1500s, and in some places it paid well and became a fairly prestigious job. Digging a grave took from four to eight hours, depending on the ground conditions and the number of diggers involved, because people had discovered that shallow graves allow odors to rise that attract insects and varmints. The gravedigger eventually became the general caretaker, but up until the late nineteenth century, he left the closing of the grave to the pallbearers. Now whoever digs the grave generally finishes the job.

In the 1950s backhoes came on the scene, considerably reducing the amount of time and effort required, but I found that kind of disappointing. I was hoping to talk to someone who had a lot of time to think about what he was doing as he dug into the ground.

Although Bennett recalled no real surprises on this job, the man who'd held it before him had had a few accidents. "You have a survey that shows you approximately where an older coffin is supposed to be," he told me, "but it's not always in that position. The ground shifts or they weren't careful about marking it on the map, and sometimes he'd hit a coffin and break it open." He arched an eyebrow to let me know that the decomposing contents made the job rather macabre.

I glanced over at the doors built into the hill and pointed. "Is that just for tools?" I asked.

Bennett smiled. "Now it is," he acknowledged, "but in older days that was the cold storage."

"Cold storage?"

"Yes, for bodies. If people died during the winter and the ground was too hard to dig up, they'd be stored till a thaw. That was the holding area."

My mouth dropped open. So it *was* as spooky as I'd always thought. I hugged myself against the sudden chill.

Bennett had to prepare for the impending funeral, so I took my leave. Yet as I walked away, I realized that, backhoe or not, gravediggers had to see things. From my background in forensic investigations, I knew that to get a body back out of a grave, one had to shovel carefully, and in the case of corpses buried for more than six months, some coffins may very well be rotted or broken.

Digging up someone who's been buried is known as an exhumation or disinterment. This can happen for many reasons, but often it's to:

- reaffirm the findings of an autopsy.

- actually do an autopsy that should have been done.

- check some detail for a forensic investigation.

- collect evidence.

- gather information for a malpractice case.

- search for a lost object.

- ensure that the body in the grave is the one that's supposed to be there.

It has even been for personal reasons, although this is uncommon. Abraham Lincoln got a gravedigger to exhume his dead son, Willy, on two different occasions so he could gaze at him again.

I spoke to one man who had been hired to exhume the coffins of a mother and two-year-old daughter who had been murdered. The investigators had believed that an intruder had come into the home and had stabbed them with a knife that was found in the backyard. The coroner had instructed the pathologist to look for anything that supported that theory, and no one bothered to take prints off the victims, get hair samples, or even to test whether the knife actually fit the wounds. Then when suspicion fell on the father and another weapon was discovered, the bodies had to be brought up for a more thorough examination. They'd been buried over six months earlier, so when the gravediggers broke open the vault that covered the coffins, they were met with an overpowering stench.

"What happens," this man told me, "is that the body fluids from decomposition rust out metal or rot the wood, and at some point the bottom of the coffin usually gives way. So you can't just hoist it out. In this case, there was some damage, but we managed to get them onto a supporting platform and get them out. I don't think I'll ever forget that smell."

So gravediggers are quite instrumental in getting back into a grave, and that means they may have to deal with a decomposed body or a skeleton. Robin Wall, a mortician in New Mexico, got involved with Tarpening & Son Mortuary in moving an entire cemetery—a "boot hill" graveyard in Seven Rivers, where men had been buried after violent deaths. In 1988 the painstaking disinterment of fifty-four separate graves began. Here and there a skeleton was found, along with bits of clothing and coffin nails. One body proved to be that of a male, five feet tall and around age twenty, with all of his teeth still intact. A few corpses had been buried without coffins, and some of the remains showed obvious signs of having been shot, such as a bullet hole through the skull.

"We laid it out in another town exactly as we found it," Wall stated. "It was a two-month ordeal."

In the event of floods that erode a cemetery, or construction that requires the transfer of graves from one place to another, gravediggers may witness a wide variety of things that happen to people once they're in the ground.

One gravedigger actually kept track of where a body was buried because he had something in mind for it. This was in Vienna, Austria, in 1791. The famous composer Wolfgang Amadeus Mozart had died at the age of thirty-five from a "fever." As was the practice then, his corpse sat in a coffin awaiting darkness, when it could be transported to the cemetery near St. Marx. There it was buried in a winding sheet in a common grave, while the coffin went back to the church to be reused. There was no marker, but gravedigger Joseph Rothmayer claimed that he'd made a point to remember the location. Then when Mozart had turned to bone and was dug up ten years later for redistribution (another common practice), Rothmayer grabbed the skull. Scraping off the last bits of flesh, he kept it for himself. After *his* death, it changed hands several times until, just over one hundred years later, it ended up at the International Mozarteum Foundation in Salzburg.

These days, gravediggers are not quite so personal with the dead, yet odd things can still happen in the process of preparing a grave—things that most of us don't want to know about. Don, a funeral director in business for eleven years, told me about one of his acquaintances who used the same vault company as he did. (Many cemeteries now require a vault or cement lining to encase the casket to prevent the grave from sinking.)

"He had to use this special oversize casket for one man," Don said, "and after the family left, they went to lower the casket down. It turned out to be too large for the vault, so the hearse driver called the funeral director and said, 'We have a problem.' He came out, took one look, and then stepped on top of the casket. It didn't budge, so he tried jumping on it to force it to fit. That didn't work, either, so he had to go back and get another casket. Of course, since he'd needed an oversize casket for this man, the body didn't fit very well into the new casket, so they ended up forcing it in. Finally they were able to lower the whole thing into the vault. People don't generally know it, but that kind of thing can happen."

It wasn't the first time I'd heard about abuse of a body for the purposes of burial. One undertaker who cared little about his clients would sell an oversize casket for a tall man, for example, and then stick him in a less-expensive regular-size casket. How did he do it? He'd just cut the feet off. He figured that the corpse would never know the difference and he'd made himself a few bucks.

I had met Don through someone who told me that he'd heard about bodies actually exploding inside caskets. I couldn't imagine why that might happen, so I asked about this. Don just laughed. "There was an ad in the paper once," he explained, "run by some religious association. They wanted to inform their people that funeral directors and private cemeteries were in cahoots with each other, so they claimed that bodies exploded in the mausoleums and body parts were thrown everywhere. It wasn't true. About all that happens is that a seal might blow out on a casket, so now they've stopped sealing caskets in mausoleums. But this group was trying to scare families. They falsely claimed that I'd be getting up in the middle of the night to clean up the mess, but I never did that."

At the time, I accepted this. It was only later, when I looked more deeply into the business of coffin manufacturing that I realized that, in fact, bodies have exploded in some coffins and it's become quite a scandal.

My conversation with Don got me thinking about just what happens when a body dies, so I looked into it and I wasn't too happy about what I discovered. While we're alive, the body acts as a system to distribute oxygen and nutrients. At death, all of this stops, and then certain changes begin happening to the corpse. According to Kenneth Iserson in the definitive *Death to Dust*, corpses go through the following stages:

1. When the heart stops beating the blood begins to settle into whatever parts of the body are closest to the ground. The skin pales and starts to look waxy.

2. In the state of livor mortis, which begins within an hour and gets fixed eight to twelve hours later, the skin discolors into a purple-red. The eyes flatten and the extremities turn blue. The body temperature has been cooling at a rate of about one to two degrees per hour, depending on the outside temperature and the amount of body fat.

3. At first the muscles relax, but then they stiffen into rigor mortis, which generally shows up first in the face and soon spreads visibly over the entire body. It can stay this way from one to three days before it relaxes again. This softening results from the muscle fibers decomposing.

4. It then putrefies, signaled by a greenish discoloration in the skin, starting in the lower abdomen. Soon this will hit the entire body, and the face will swell and become unrecognizable.

5. As putrefaction spreads, a foul odor develops. Bacteria in the intestines produce gases that bloat the body and eventually turn the skin black. The bloating also makes the tongue and eyes protrude, and pushes the intestines out through the rectum.

6. Soon the skin blisters from these gases, detaches, and bursts, and then the internal organs break open and liquefy. At this point, the body starts leaking from all the orifices.

It's no wonder we want to get these bodies out of our sight. Embalming or refrigeration delays this process and cremation eliminates the most harrowing stages, but it's still important to get right to work soon after death occurs.

Yet things haven't always been fastidious. In the Middle Ages, cities became overpopulated and burial grounds quickly reached capacity, and then some. Then epidemics swept through Europe, which only made things worse. Some people were buried in the church floors, but others were tossed onto trash heaps or stacked against a church building. At that time cremation was banned, so there was nothing to be done but let these bodies lie around and rot. Some piled up inside a church while others were actually laid out on the roof. Unfortunately, as they bloated and the skin burst, they leaked infectious fluids that made the general area hazardous. If they weren't buried, they eventually liquefied right where they lay, to the disgust of those who had to pass by. The odors were unimaginable.

So now bodies generally go into the ground, get cremated, or go into mausoleum vaults. What happens there?

Bodies decay in response to various factors:

1. Warm temperatures accelerate decay, as does exposure to insects, and an exposed body can become skin and bones in a couple of weeks.

2. Burial without a coffin will reduce the body to a skeleton in about ten to twelve years.

3. Coffins can slow the decomposition down considerably, and some coffins opened a century or more after burial show the remains still intact—even in relatively good condition.

4. Dry, hot climates can mummify the skin and organs.

5. Bodies immersed in boglike conditions may be preserved indefinitely. Corpses of bog people can be five thousand years old.

6. A thorough embalming can slow decay for centuries, although most people only get a partial embalming.

These days, funeral directors take great pains to protect the public from seeing how disturbing the various stages of death can be. They take a corpse behind closed doors to make it presentable for viewing and burial.

Funeral Homes

My first introduction to funeral directors was through scary movies like *The Comedy of Terrors*. Imagine Vincent Price, Peter Lorre, and Boris Karloff all in the same film about undertakers who beef up their business in a rather unscrupulous way. Karloff is an aging undertaker who hands off his funeral business to his son-in-law, played by Vincent Price, whose assistant is Peter Lorre. To get "customers," these two decide to break into the homes of old men to kill them. The gaunt Price and the short and fumbling Lorre make for quite the creepy team, and thanks to Hollywood, I had generally imagined the prototypical undertaker as being like them: secretive, detached, and somewhat warped. Hence it was with some trepidation that I first approached a funeral home.

The tradition of the funeral home dates back to the Civil War, when buildings were needed for embalming the bodies of soldiers so they could be returned to their hometowns. At first it was the coffinmaker who provided this service, but increasing numbers of people in urban areas wanted to enlist the services of a professional to take care of the business of death. A place to hold wakes was needed, so funeral parlors provided special rooms. Eventually, the undertaker took on more of the funeral-related services, and these days, they pretty much offer to take care of everything that needs to be done.

According to the National Funeral Directors Association (NFDA), a funeral director manages all aspects of a funeral service. That means that they pick up the body, supervise its preparation, get the death certificate, sell a casket, arrange for service personnel such as drivers or pallbearers, orchestrate a memorial service, get people (and the body) to the cemetery (or crematory), talk with clergy, and make sure that the grave gets properly closed. They may also suggest grief counselors and follow up on the surviving family members.

It was the Frank E. Campbell funeral home at 1076 Madison Avenue on New York's Upper East Side, where services have been performed for John Lennon, Bat Masterson, Jacqueline Kennedy Onassis, and James Cagney, that appeared to have the most significant impact on the public acceptance of the funeral home. As Manhattan grew more crowded and people had less space in which to hold a wake, this spacious funeral home offered ways for people to lay out the body of a relative and invite as many participants as they desired.

Back in 1898 it was called the Frank E. Campbell Burial and Cremation Company, but as Campbell became more involved with grieving families, he saw the need to take over many of the details. Thus, he founded "The Funeral Church," in which he offered to provide an atmosphere of serenity and beauty. However, behind the scenes he also orchestrated public displays of grieving. For example, when silent film star Rudolph Valentino died from an ulcer at the age of thirty-one, the Frank E. Campbell Funeral home hired people to create a sensational atmosphere among the mourners (some of whom he'd even hired to "grieve"). He got a lot of publicity, and afterward it became more common to see a boisterous funeral procession coming to his place. Thanks to Campbell's innovations, traditions that once had been deeply private became ever more public, and his own reputation as *the* funeral director for the rich and famous was enhanced.

For many generations, the funeral director and his family lived on the premises. These days, one is just as likely to find mortuary students or caretakers in the place as the family. In the past decade, large corporations like Service Corporation International have been buying up the independent funeral homes and then clustering them to cut costs, but the "independents" resist this, feeling that personal service from a local citizen provides a better experience for people enduring a loss. What I hoped to do was get inside a funeral home where people still lived and see what that was like.

I got that opportunity in Ohio. In the small town of Leetonia I spotted an impressive white mansion that turned out to be the Wood-Rettig funeral home. It didn't look like a place where Vincent Price might live, but one can never be sure. I went in and introduced myself to Nettie and Wallace Rettig, and much to my relief, they weren't anything like Vincent Price. Or Peter Lorre.

Wearing a turquoise baseball cap and a heather-gray T-shirt emblazoned with LEETONIA SPORTSMEN'S CLUB, Wally invited me into a small office to tell me about the building, which was a rather impressive mansion for such a small town. Wally had gotten into the business through his son, John, who currently owns three area funeral homes. That meant that Wally had the distinction of becoming the first practitioner in the state to have apprenticed with his son.

The business was started in 1890, initially renting space from a furniture store (which was typical of many funeral businesses because the furniture maker made the coffins, too). It then moved into its own building, part of which had been the carriage house for a mansion more than a block away. The upstairs was originally used as the casket-display room, so the caskets had to be taken up on a special elevator. I went inside this lift and could just imagine the children in the family riding up and down, fantasizing about corpses and ghosts. Now this floor was the Rettigs' spacious apartment. For the most part, Nettie told me, her friends and relatives accepted it, but her sister refused to sleep there.

Downstairs was the chapel, which had been cleverly designed to serve as a display for the caskets when not otherwise in use. If a service was scheduled, the caskets disappeared behind curtains drawn over the niches, so as you sat there listening to a speech about the dearly departed, you wouldn't realize that all around you were beds for the dead.

Wally said that most embalmers tended to overuse cosmetics on corpses, which gave them that pancake appearance, so he had found a way to avoid that and he offered to show it to me. We stepped over to where one of the walnut coffins sat on a stand and he went behind a curtain.

"The way we get the skin to look natural," he said, "is with lights." He then demonstrated the different effects on the coffin from the overhead red, yellow, and blue lights. I thought that was rather innovative.

Then he showed me the embalming room, which looked like a surgical center. When barber-surgeons were charged with this duty in nineteenth-century England, embalming was called "ripping the corpse," but today things are more genteel. Mostly what happens in these rooms is hidden from the public, so relatives of the deceased will only see the finished product.

Since they had stopped using the room for embalming, I was legally allowed inside. A year earlier, Wally explained, they had transferred all of that to another home, but the steel drainage table and sink were still in place. In the cupboards on the side, I could see bottles of chemicals and powders, and I knew there was also a storage place for the various instruments—scalpels, clamps, forceps, needles, and tubes. There were also props, like head rests, to stabilize the corpse.

As we left the prep room, Nettie told me that she had been a beautician and had become pretty good at styling hair for those who had passed on. I hadn't thought of this, but in fact many beauticians include among their skills styling the hair of the deceased. Nettie's first job had been to do the hair for her boyfriend's dead grandmother, and at the time, she'd been nervous enough to ask that someone else be present. However, she got used to it, just as she grew used to living in the same building where corpses awaited preparation and transport.

While it was great to get the tour, I wondered what it must be like to actually grow up in such a place. To my mind, it had to be strange.

Blair Murphy, the son of two funeral directors, was willing to tell me something about that. "When I was a little kid," he said, "the school asked each of us to write what our fathers did for a living. I wrote 'My dad paints dead people's lips,' and it got printed in the town paper. That was my first press quote."

A filmmaker, he'd produced *Jugular Wine* and *Black Pearls*, in which he expressed his perspective on the attraction to spirituality and death. Since his parents were both in the death-care business and his father was the county coroner, discussions about corpses around the dinner table were commonplace. Thus, Blair was exposed early to macabre subjects.

Blair, who is in his mid-thirties, has traveled extensively and now lives in California. Meeting him was rather disarming. Despite a closely shaved head that indicated a certain degree of social rebellion, his face was rather wide-eyed and cherubic. To see his films, I accompanied him to his apartment one October afternoon.

I was struck right away with the museumlike aura. On every wall and in every corner I could see the evidence of Blair's trips to places such as Egypt and New Orleans. I looked around at the exotic paintings, Sphinx tapestries, and framed poster of King Tutankhamen's golden burial mask. From his father, Blair had gotten a casket with a glass top that he used for a coffee table. His mother had given him a stuffed coyote to enhance his taxidermy collection of "low-maintenance pets."

Gesturing around the room, he smiled and asked, "Where do you think these things come from?" Before I could answer, he said, "Obviously having been raised among the dead has influenced me. And none of it's spooky to me. I just think of it all as beautiful . . . like the gleam of gold by candlelight in the night."

He invited me to sit down, and I soon learned that his father had been a funeral director for almost forty years. It was all the man had ever wanted to be, and he'd hoped that his son would follow suit. Like many an undertaker's son before him, Blair had helped place bodies into coffins, but ultimately the funeral trade had not gripped him.

"Being raised in a funeral home never seemed odd to me as a child," he commented. "In my early teens, it got to be a drag because it was such a conservative business. I was always being asked to smile and look nice for cameras when, really, I was ungodly angry and rebellious. All I wanted to do was move to Los Angeles and become degenerate."

"But you don't seem bothered by it now," I pointed out.

"By my late teens I thought it was actually pretty cool and even a great honor to have been raised in such a rare environment as a funeral home."

"So what about how you lived?" I pressed. "Wasn't it strange to know there were bodies in the same building where you were sleeping?"

"The funeral-home aspect was just a house," he responded, "but embalming and being around dead people was interesting. I can remember being alone often with dead people and staring for long periods of time into the partially cracked eyelids. I dared myself to see how long I could stare, with the thought that if their eyes moved even a speck, it would forever alter my sense of reality. That was kind of thrilling."

Blair mentioned that he had a sister, Donna, who was affected differently. "She doesn't feel the same way about death that I do, but we've both had many dreams where dead bodies wake up and speak to us or chase us or confront us."

Rather than tell me her story, he introduced me to her, so I was able to learn about growing up in a funeral home from both perspectives, male and female, brother and sister. Nineteen months separates them, and Donna feels that she's "all business." It was her father's position as coroner that most influenced her, so after college she became a private investigator.

"The job of a coroner was often high profile," she said. "It wasn't unusual to see my father on television or in the newspapers. As a child, I took great pride when I saw him being interviewed about a recent unattended death. When I knew that he would be called into a case, I would comb the newspapers and absorb everything I could get my hands on about the case."

She agrees with Blair that they could not have turned out to be more different. While he ponders reincarnation and accepts a person's demise, Donna feels that people who die have paid the ultimate price. "I cry for the dead. I honor them by attending their funerals. I feel a genuine sense of loss."

While growing up in that environment, certain experiences made a deep impression, in particular the local murder of a child.

"There was this nine-year-old girl," Donna recalled. "She had been sexually assaulted and murdered by a twenty-something-year-old neighbor. For days, the girl's disappearance had been a top news story, and in typical media fashion, the details of her life had been published at length. As an eight-year-old, I was so fascinated by the story that I read every account available. I remember reading about her favorite games and toys, about her family and about the time leading up to her disappearance. I read so much that I felt as if I'd actually known her. Then when her body was discovered about a week later, my father was called on as the coroner to retrieve it and then later, as funeral director. I couldn't believe that this girl, who I knew so much about and whose death had been so publicly reported, was being brought back to our house for preparation.

"When she was finally dressed and laid out, I recall that I stood over her casket for a very long time. This was the first time that I had ever seen a small white casket and I thought it was appropriate for her. She looked beautiful in her pale pink dress and matching painted fingernails. But I was also deeply afraid. I remember thinking that I did not want to turn nine because I was scared that such a gruesome demise would happen to me."

It's clear that Donna will continue to be fascinated with crime, while Blair expects to move more deeply into innovative projects that involve themes of death. He's even set up his own Web site, TheCemetery.net, to put them on display. "I'm delighted to have this mortuary background naturally in me," Blair said. "It's created me. And now I'm just taking certain aspects of this culture in a different direction."

For me, it was quite interesting to see how two children from the same death-care-centered environment had developed such contrasting, yet complementary lifestyles.

It was the childhood experiences of a woman named Cynthia that most reinforced my belief that there had to be something unusual about being the child of an undertaker. I'd talked with a lot of people who'd been raised in such an environment and had been disappointed that most of them just took it in stride. Cynthia did, too, but clearly her experience was unique.

Her father loved the business so much that he felt his four daughters ought to have a thorough exposure to every detail of it. On his days off, rather than rest, he took them on educational field trips—to the mortuary.

"A mortuary visit always included viewing the dead," Cynthia said. "So my father took us into chapels in which bodies had been laid out, and he would comment on the hair, the makeup, and other cosmetic fine tuning. He could make most people look better dead than they ever did alive, and he took pride in that. He said it made people who've lost someone feel better."

Then they went to where the bodies were prepared. "We always went to the embalming room. We would stare at naked bodies plopped down on slabs, and sometimes a body was so corpulent it would hang over the sides of the table, but usually the dead were withered to skin and bones." If an autopsy was to be performed, Cynthia's father made a special trip to show them. "He never wanted us to miss a full dissection." Although such access was forbidden by law, "the preparers of the dead abide by their own code."

It was during the dark hours of the night that Cynthia learned about "the call"—the trip to pick up the body when someone had died. "Night calls were when my father had free reign of the mortuary and he used those wee hours to complete our indoctrination. When someone dies in the middle of the night, usually the hospital or home wants to get rid of the body pronto. So the morticians take turns. My father would allow one of us to go with him. Although he would slip into a shirt and slacks, whichever little girl was riding shotgun would still be in pajamas."

He'd exchange his car for a hearse, and then drive to the removal location. He'd get out and then return with the body on a gurney. After loading it in the back, he'd drive to the mortuary, and that's when the interesting stuff happened. "The body would sometimes moan or sigh or whistle. Gases leaked noisily from orifices or an enthusiastic limb might thump against the floor. Even my father would be a bit spooked."

Since it was tough to raise four girls on his wages, Cynthia's father would collect whatever clothing a grieving family did not want and bring it home. "Most of our nightgowns and bathrobes were previously owned by the dead," she said.

Apparently this did not bother her, but I'm not sure that I could have worn such clothing.

Being familiar now with the inside of a funeral home, I was curious about how the dead actually got there.

The Call

"That's how it starts," one undertaker told me. "You get a call that someone has died. It can happen at any hour of the day or night, and you can see some pretty strange things at removals."

Upon the discovery of a death, a call goes out to 911 or the family physician. If the person died unattended and was discovered later, a professional must come in to pronounce the person dead and record the time. That means a police officer, paramedic, nurse, or physician. Even at an accident scene, the removal of bodies requires some sort of official release. The body may then be sent to a funeral home, a hospital, or a morgue. If sent to a morgue, someone must come in to identify the deceased and then the corpse's "John Doe" toe tag gets a name. With undetermined deaths that seem suspicious, a coroner or medical examiner may be called in to decide if a partial or complete autopsy is required. If so, the body remains at the morgue. Generally, funeral directors hired by the family pick the body up from there.

Many people think that coroners and medical examiners are pretty much one and the same, and that's true in some states, but in many other states, there are significant differences. A few states still rely on the coroner system for medico-legal investigations, but not all coroners are pathologists and they're generally elected to the job. Some may even be funeral directors. On the other hand, medical examiners are trained as forensic pathologists and they're appointed. Their job is to perform the autopsy and give a medical opinion.

Whatever the case is, people who arrive at a death scene for a removal have to be prepared for anything. I heard of one situation where the police were called to investigate an overpowering odor in an abandoned building. They went in and found a male corpse lying spread-eagled and facedown on the floor of an apartment. He was in pretty bad shape; his soft tissues had decayed to the point that his skin, from his face to his lower torso, had glued him to the spot. They tried to pick him up, but he didn't budge. They had to call for help and actually use shovels to finally pry him loose.

Another case involved a woman who went about cleaning her home and organizing her affairs, washed and ironed her clothes, bought a new dress and got her hair styled before she hacked herself to death with a machete. No one could make much sense of that, or why she had chosen such a brutal way to end her life.

An ambulance driver called to one scene had to go with her partner down an unusually narrow and winding set of stairs to pick up an eighty-year-old man who'd been dead for several days. While they got the gurney down the steps by turning it sideways, they knew they'd be facing a problem going back up. The man was lying in a bed, and the small, sparsely-furnished room was warm, so he was decomposing. Breathing through their mouths to stave off the odors, they lifted the corpse onto the gurney and strapped him in. Then they proceeded toward the narrow staircase. However, it soon became apparent that this was not going to work. To get him up the steps, they'd have to turn the gurney in such a way as to dangle him by the straps, so they went back into the room to think up a better solution.

For a few moments, it seemed like they had an impossible task. There was just no way to get this man out. Then they had an idea. Opening the small basement window, they realized they could pass him through there, so they wrapped him in a sheet and one of them went outside to grab hold of the other end. By shoving and pushing the sheet-wrapped corpse, they finally got him out.

Since death is not predictable, funeral directors are on call all the time. The call generally prompts immediate action—even when the person is not yet dead.

Funeral director Daniel Hartzler grew up in an area where few phones existed, so people relied on a system of communication to get news. One family had a relative in the hospital and they heard through the grapevine that she had died. The natural thing, then, was to show up at the funeral home. "They came in and rang the doorbell, and my father met with them to make the preliminary arrangements for this woman. Then he called the hospital and learned that she was still alive. He relayed this news to the family and they were shocked, so they left. Two days later, they came back and rang the doorbell. When my father answered it, they asked, 'Is she dead this time?'"

Many funeral directors dread the obese person who might

have to be taken up or down flights of stairs, and Hartzler had one such case. "We had a large lady who had died on the commode. The bathroom was very small and we had trouble getting her out, but finally we did. We laid her on the stretcher and covered her as quickly as we could. Then we got her back to the prep room and put her on the table. Oddly enough, we would just touch her right shoulder and she'd roll to the right. Then we'd touch the other shoulder and she'd roll over to the left. So we wondered what in the world was going on. I got down and looked, and found that here we'd pulled the commode seat off with her. It was still stuck to her."

It also happens that someone is found in such an odd position that no one quite knows what to do, and it turns out that the death was a rather embarrassing sort of accident—the result of dangerous self-arousing sex. One thirty-five-year-old man was found dead in his home, lying on his stomach and trussed up in a rather astonishing array of devices. He'd been dead for several days, asphyxiated by a black plastic bag placed over his head. His feet were tied together and raised up behind him by a rope that was attached to a metal hook in the ceiling. His arms were likewise bound behind him with a series of belts. Supposedly he had tied himself up this way many times without incident, but this time his device had finally gotten him. Part of the achievement of ultimate stimulation for people such as this, it seems, is the potential for death. They get close by temporarily limiting the oxygen supply.

Roy Hazelwood, a former FBI special agent, is one of the premier experts in the country on autoerotic fatalities (also known as terminal sex), most of which involve males. Of the hundreds of deaths that he has studied, the average age was twenty-six, but the oldest was seventy-seven and the youngest he knew about was six. Sometimes it's the case, he points out, that a "teenage suicide epidemic" has more to do with passing around clandestine "secrets" than with psychological contagion.

People who die this way may use ropes, plastic bags, electrocution, aerosol propellants, and mood-heightening drugs. They might be wearing anything from a horse bridle to cross-gender underwear to cowboy outfits, or nothing at all. Many use bondage equipment.

Hazelwood reports the discovery of a young man found in a garbage can. He had trussed his ankles and wrists with roller-skate straps and then lowered himself into the can to try to induce a state of hypoxia, or lack of oxygen. His escape mechanism consisted of a heavy roll of wire that sat next to the can to help him tip the can over and get out. However, he underestimated how far into the can he'd settle, and it's likely that his death was long and quite agonizing—not to mention humiliating to his family. He apparently did call out for help, but people in the neighborhood took it for a howling dog.

Sometimes people go away from home to practice their clandestine activities, which means that if they end up dead, some stranger will find them. A fisherman walking into the turn-around area of a secluded road came across a naked man chained to a car, crushed under the left rear fender. The left door was open, the motor was running, the transmission was in low gear, and the steering wheel was rigged to make a sharp left turn. Tracks in the area indicated that the car had made numerous circles. The police arrived but couldn't figure out what to make of this scene. Only an experienced medical examiner finally deduced what had happened.

The man, a forty-nine-year-old married father of two, had fashioned a chain harness for himself that looped around his neck and went down to his waist and between his legs to fasten again behind him. Then another chain attached him to the car's rear axle. To all appearances, the man got sexually excited by being dragged behind the car, and so this is the way he'd made the scenario work. However, in this instance the chain must have gotten caught in the wrong way, because it was wrapped up around the rear axle several times and the car had rolled onto him. His clothing was found in a gym bag in the car, along with the tools he'd used to make the harness. However, the story he'd told his wife was that he was going out for some target practice.

Not everyone who gets removed is actually dead, and that can really make people nervous. In Ashland, Massachusetts, on January 25, 2001, paramedics answered a call to find a thirty-nine-year-old woman sitting in a tub of cold water, an apparent suicide from a drug overdose. They detected no signs of life in the bluish, stiff body, so they called the medical examiner to describe the situation. He told them, "Take her to the funeral home." He would look into it later.

They zipped her into a body bag and took her by ambulance to the funeral home where John Matarese was to prepare her. However, he was on his way out for lunch, so he asked that she be placed in the holding room. The paramedics did so and left. Then, as Matarese walked past that area, something caught his attention, so he stopped to listen. He thought he heard the sound of breathing. Stepping inside to check, he realized that a gurgling noise was coming from inside the body bag holding the most recent "client." Shocked, he unzipped the bag.

"It scared me half to death," he said later. "The girl was alive." Calling on another team of emergency personnel, he got her transferred to a hospital. Since this became national news, Massachusetts declared that it would look into its current procedures for the transfer of a corpse.

An Autopsy

People who die unattended deaths where the cause cannot easily be determined are supposed to be evaluated for an autopsy, although sometimes the family will oppose that. Those destined to get one for legal purposes end up in the morgue and the others remain in the hospital for limited testing or get removed to the funeral home.

Now, the morgue has an interesting history. The concept started in the 1700s with the high number of street murders in Paris. To get some of the bodies identified, police stuck them in a vacant butcher shop called *Le Chatelet,* so that people who knew the victims might see and claim them. These days morgues are holding areas for refrigerated corpses while their fate is decided. They can stay in the drawers for about four days or so without suffering the effects of decomposition.

When funeral-home personnel come for a "removal," the morgue attendant is responsible for locating the right drawer in which the deceased is kept, and he's supposed to check this on the toe tag. Yet I heard about a person with an amputated leg who was taken for a closed-casket burial, when in fact the man they should have taken had both legs. Somehow the family found out and they had to have the guy dug up and replaced with their own relative. Of course they sued the funeral home, but the removal team pointed the finger at the morgue attendants for not reading the tag, and they in turn pointed it back at the team.

A medical autopsy is a postmortem examination of a corpse to determine cause of death for an official report. Around 25 percent of all deaths in the United States are subject to it. If the death is suspected to have been part of a crime, the procedure is known as a medico-legal autopsy, which is done for the courts. It's routine in homicide cases. The autopsy itself can be partial, selective, or complete, but most are partial. That means that only part of the body is examined for cause of death. A selective autopsy may only involve one specific organ, such as the heart or brain.

Autopsies are generally performed in a sterile environment, although I read about one that took place at a grave site. This was in 1662 in Hartford, Connecticut. An eight-year-old girl had died mysteriously and the townspeople wanted to determine if this had been caused by a supernatural agency, such as a witch might use. The physician who performed it found the girl's throat wholly contracted so that nothing could be forced through, and he could not determine any natural cause. Rather than await her fate, the suspected "witch" left town.

As part of a research assignment (not for witchcraft), I once witnessed an autopsy in progress. As I came into the sterile medical lab, I noticed bodies being prepared on several steel tables with holes for drainage. Two had already been cut open, including an obese woman, and the sight of her layers of fat reminded me of a whale. These people just didn't look human.

I turned away to where the medical examiner was removing the organs from a middle-aged man. Over his table was a smaller dissection table for cutting up and examining organs, like the liver, and nearby was a hanging scale for weighing the organs as they were removed. A large tank on the floor collected fluids. To one side were sinks and pieces of equipment such as circular saws, and along one wall were shelves containing jars of liquid in which preserved specimens were floating.

I hugged myself against the chill in the room. I had been advised to breathe through my mouth to diminish the intensity of the stench of decomposition and blood, although I actually didn't notice that it was that bad. What *was* bad was seeing a corpse slit open, with a pile of organs on his chest. At this stage,

I realized that the body had already been photographed, X-rayed, weighed, and measured, and had been opened with what's called a "Y" incision. That's a cut from shoulder to shoulder and then straight down the abdomen to the groin area. I'm glad I hadn't seen the knife go in.

"So what does an autopsy tell you?" I asked one of the officials. "Besides the cause of death."

"We can figure out just what went wrong in the body to cause it," he said, "such as a brain hemorrhage or asphyxia, and sometimes whether the death was accidental or not. Anyway, we can give a medical opinion on it. For example, we can say from the condition of the lungs if someone died before being dumped in a lake or if she died because she drowned. We might also be able to determine the type of weapon used, as well as find clear evidence of sexual assault."

I saw a scary-looking saw, stained with blood and tissue, sitting on a counter by a sink and I learned that it was used to cut open the ribs to remove them in one piece in order to get to the internal organs.

"We then take a blood sample from the heart to determine the type, and then start taking things out to weigh them. If there's fluid, we aspirate it to get a sample, and we also open up the stomach to get a look at the digested contents."

"Then what do you do with it all?"

"We make slides for further analysis and if they need any of the organs for evidence, we have to preserve them. Otherwise, they go into a bag and get put back into the body before it's sent to the funeral home."

To sum it up in simple terms, an autopsy follows these steps:

1. The chest comes first. The medical examiner goes straight to the heart and lungs, which are removed and weighed.

2. The abdominal organs are examined and removed, then weighed.

3. The genital area gets examined for injury. Swabs are taken from all orifices.

4. Blood, semen, and hair are collected for DNA typing.

5. Urine is removed and sent for drug testing.

6. Finally, the head is examined.

For me, this was the worst part. I watched while the eyes were probed for hemorrhages that could reveal strangulation. After that, an incision was made in the scalp at the back of the head and I forced myself to watch as the skin was carefully peeled forward over the face to expose the skull. As bad as this was, I knew there was more to come that involved the saw. Now I was breathing through my mouth, and I watched as the high-speed oscillating power saw was placed against the skull. A loud screechy *whirr* told me they were cutting through the bone. (It reminded me of the movie *Hannibal*, where the top of a man's skull is removed while he's conscious, exposing the top of his brain.) They then used a chisel to pry off the skullcap.

Once the brain was lifted out and placed on the table, there wasn't much reason to wait around. Organs would be sliced up for slides and the rest of the procedure would involve other types of analyses. I left the room just as they were wheeling in a dead woman. Her eyes were wide open, as was her mouth, and her rigor-stiffened arms rose into the air as if she were trying to catch something. I skirted this gurney and wondered why she was in such a pose.

"Unattended death," the official commented. He seemed rather jaded about these things. "She probably died in her sleep, on her stomach, with her arms stretched out by her head."

If there's nothing left to do for the autopsy, the funeral home designated by the family comes in to remove the body.

◆ *Preparing the Corpse*

According to many funeral directors, there are more than one hundred decisions to make in the process of planning a funeral. From products to services to grave preparation, handling a death is not as easy as it was in the days when they just wrapped a corpse in a sheet and lowered it into a hole.

One surprise for me was the discovery of "burial societies." The one I heard about observed the Jewish ritual of washing and preparing a corpse for burial. With embalming forbidden, this all gets done quickly, but not by the mortician. Volunteers from the community care for the dead as a religious duty. Two or three people of the same gender as the corpse go to the funeral home to take care of details. To proceed with the purification ritual, they remove everything from the body. First they say a prayer, which continues while the body is washed—but never turned facedown. Then they pour three bucketsful of water over it, saying, "He (or she) is pure." Some then put a spot of egg white on the forehead (believed to have set apart the Jewish corpses during the European plagues). In the final step, they clothe the body in a linen shroud, with a hood or veil that covers the face, and place it in a wooden casket that has no metal on it, not even nails.

Most other people just get worked on by the undertaker.

I like the word *undertaker*, which refers to someone who undertakes the job of making the funeral arrangements. Such a job became "official" when someone in town basically undertook to make arrangements for other families besides his own. However

I soon learned that many people who do this job as a profession dislike being called undertakers.

In the 1880s they officially began referring to themselves as funeral directors, although many were really furniture makers who ended up doing funerals on the side because they supplied the coffins. The idea was to move them from the "craftsman" category and into that of "professional." Within a few decades, they used the word *mortician*. Some even tried *doctor of grief*, but that didn't catch on, and finally the appropriate title is now funeral director. (I still like to think of them as undertakers.)

I should point out here that a funeral director is not necessarily an embalmer. While some do embalm, others merely supervise the embalmer as one of their staff members, and some people just do embalming without being funeral directors.

There have been several exposés of the funeral trade—notably Jessica Mitford's biting *The American Way of Death* and Darryl J. Roberts's *Profits of Death*. That made things tough for me, because those books tended to make readers perceive funeral directors as unscrupulous practitioners, and those who truly love their work resented being so depicted. Some would only talk under terms of anonymity.

One undertaker told me about a case that drew him into a world of traditions that he had never witnessed before. The body of an older woman was brought in one day with her decapitated head sitting on top of her body. "You don't usually see it that way," he pointed out. "It was just sitting there on her chest and it was kind of a shock." He learned that she had been a Gypsy.

It seems that her killer, a large black man with whom she had been involved, had decided that she had cursed him, so he'd come into her home swinging a huge sword. He cut off her head and then used a knife to hack at other parts of her until he was certain that she was dead. Someone found her and the coroner took her for an autopsy. After that, she was transported to the funeral home to be prepared for a viewing.

Some funeral directors might have opted for a closed-casket funeral, but this woman had been high-ranking royalty, which meant that many of her people would attend the funeral expecting to see her. So the undertaker went to work.

"We didn't want them to have any reminders of the incident," he said, "and she had lacerations to her face and hands, so we had to stitch those up and use wax and makeup to cover them. We got some metal dowels from a lumberyard to reattach the head—in the old days they might have used a broomstick. Then the wounded areas had to be seared with chemicals before the head was attached. After that we sutured the muscles and the skin, and seared the areas again. Next we applied a special adhesive to make sure that everything would stick together, and used cotton and plastic to prevent seepage from the wounds. That way we could carry and maneuver the body without the head coming off. We dressed her in a high-necked white sequined dress, with jewels, and applied makeup. When it was over, she looked great. Even her family was impressed.

"The Gypsies then had a party. They rented a tent for the funeral-home parking lot, roasted a pig during the service, and had a picnic in the cemetery. And her curse apparently worked, because her killer ran out to the tracks and jumped in front of a train."

Embalming Secrets

Among the members of the Electronic Funeral Service Association (EFSA), I encountered a female funeral director named Erica. She invited me to come and meet her, but mysteriously, she would not reveal her last name. She wanted to save that.

Erica lived next to the funeral home where she worked, so I drove there and knocked on her door. The woman who opened it was tall, with curly brown hair that fell past her shoulders, heavy black lining around brown eyes that accented her pretty face, and long manicured nails painted with multiple designs. When she saw me looking at her hands, she said, "I cannot embalm without my nails. I can't function with short nails."

Her name, it turned out, was Erica M. La Morte, which in Italian means "The Dead." She wore a black suit with trousers and a lab-coat-type jacket over a white shirt. Inviting me in, she pointed me toward a great room that contained a living area, dining area, and kitchen. I was struck right away by the abundance of flowers. There were bouquets in vases on the tables and more hanging over the kitchen sink. I saw flowered calendars and photos of flowers on the wall—even the picture frames had roses on them. I wondered if there might be an open coffin somewhere in there, but Erica just loved flowers.

Erica, who was twenty-five, was quite happy to tell me about herself and to teach me whatever I wanted to know about the funeral trade. "I decided I wanted to be a funeral director when I was sixteen," she explained, "after I went to the wake of my ex-boyfriend's grandfather. I don't know why this affected me, but when I walked out, I knew what I wanted to do."

Erica also had another link to the trade: "I think my great-grandmother used to help wash the dead people. This was in the early 1900s. She would go to their houses and wash their hair."

She was referring to the way bodies used to be prepared for burial by "layers of the dead." (Not layers, as in piling them up, but people who laid them out.) In the nineteenth century, people would often lay out a body at home, where women washed it, plugged the orifices, closed the eyes and mouth, and dressed it for burial (similar to the procedures of a modern Jewish burial society). They placed coins on the eyes to keep them closed, because the open eyes of the dead were considered bad luck. For a while, bodies were simply clothed in shrouds, but eventually the idea that corpses should look their best as they enter eternity became part of the modern Christian ritual.

Reminded of the part women once played, I asked Erica about women in the business and was surprised to learn that nearly half of the students in her mortuary-school classes were female. That's quite a change from even a decade ago, when men dominated the business. In the early seventies, only 5 percent of funeral directors were female.

"More women are going into it," Erica said. "They've actually been doing it for a long time, but they weren't in the public eye. They were behind the scenes, helping out. I think it's been in the last twenty years that women have become more interested in getting their license. But there were girls in my class who had never set foot in a funeral home. You shouldn't do that. You have to know that you can handle it. One girl was this petite little lady. How is she going to make a removal for someone who's three hundred pounds?"

At this point, Erica pulled out a manual that her teacher had written on embalming so she could show me what she had learned. It began with a description of Anubis, the Egyptian embalming deity.

"You have to know Anubis," Erica insisted. "He had the head of a jackal and was believed to be a scavenger who roamed graveyards. He was the collector of hearts because it was the seat of intelligence, and he removed all organs thought to decay. The 'worthless' brain was removed with a hook through the nose and discarded."

The first book to offer instructions in this once-secretive art was *The Undertaker's Manual* in 1878, and soon there were schools for teaching it. In today's mortuary schools, students take courses in anatomy, pathology, bacteria, restorative art, and physiology.

Embalming is not widely practiced around the world, but many people erroneously believe that it's a law in our country that all corpses destined for burial must be embalmed. Some funeral directors even insist on it before a cremation. In fact, bodies must only be embalmed to be transported elsewhere for burial, for an extended viewing period, or in the case of certain infectious diseases. (Each state has its own regulations.) Jews never embalm, and they get their dead buried quickly.

Embalming dates back to the ancient Egyptians and means disinfecting the body to reduce the presence of bacteria and slow down decomposition. This helps to restore the appearance of the deceased for open-casket viewing. Originally, embalming meant removing the internal organs, packing the body cavity with chemicals, and allowing the corpse to dry out. Today, embalmers replace the fluids with preserving chemicals.

In early America, people avoided embalming because it seemed a defilement of the corpse, but during the American Civil War, embalmers took on a significant role. Until that point, methods to delay decomposition mostly involved cooling, and embalming was reserved for medical specimens. Formaldehyde was not discovered until 1867, after the war, but arsenic had proved to be an effective preservative of human tissue. The downside was that it could harm or kill the embalmer. Generally, the procedure was left in the hands of the medical profession.

Thomas Holmes, a surgeon and coroner, did his own research and developed an embalming fluid that worked safely. He taught his method and sold his special fluid to undertakers trying to preserve the war dead for distant burial. Some undertakers took advantage of it to display preserved corpses as an advertisement, but this was quickly discouraged.

To transport bodies from the battleground back to their homes, the bodies were drained of their fluids and divested of their organs. Especially during the summer months, this was quite an ordeal, and embalming significantly enhanced the preservation of the corpse.

In fact, after Abraham Lincoln was assassinated in 1865, his body was embalmed. It was then placed in a casket for a two-week, seventeen-hundred-mile train ride (drawn by thirteen locomotives) to Springfield, Illinois. Not only did it survive that, but the train also stopped in each town along the way so officials could open the coffin for the people to have a look. Some seven million people viewed the preserved corpse, which influenced how embalming took hold in this country.

I looked through Erica's book to get a sense of the actual procedure, since only relatives or trainees may legally witness an embalming procedure. I had seen the stainless steel table and had learned how it drains fluid away. Now I discovered that on that table, the embalmer disinfects the body and relieves rigor mortis through limb massage. The head remains elevated to prevent discoloration of the neck and face, and nasal suction removes some of the fluids.

Since most viewings are partial—from the waist up—embalmers concentrate on the face and hands. Apparently certain things present a problem, such as buckteeth and the fact that most faces are not perfectly symmetrical. Some embalmers use cotton padding in the checks for fullness, although it can absorb a lot of moisture, but the task at hand is to make the corpse appear as close as possible to a person who's only asleep. That means finding ways to reshape sunken features. For example, eye caps (plastic forms) go under the lids, which are then sewed or glued shut. (Many embalmers prefer Superglue for this procedure.)

"We pack the throat with cotton," Erica pointed out, "to prevent fluids from coming up."

She might then insert a plastic mouth former (for a natural shape) and sew shut the lips, although this can wait until the end of the procedure as well. Before the embalmer goes to work on the rest of the body, the face has to be covered with a cloth to prevent the corpse from exhaling microorganisms when moved.

"There are four basic methods of embalming," Erica explained, "but the one most often used is arterial embalming." She pointed to the description in her book.

1. The arterial method puts the preservative into the blood vessels.

2. The cavity method injects it into the abdomen and chest.

3. The hypodermic method places it under certain areas of the skin.

4. The surface method applies creams or gels to the external body surface.

To get this done without a lot of mess, body orifices are packed and the body is dressed in plastic clothing. Then the axillary artery and vein are located for inserting a drainage tube to remove blood. After that, a decision is made about where on the body to inject the embalming fluid. Different places are selected for different purposes and according to different body types.

In fact, embalmers actually classify bodies as they come in, because that helps to determine how they will prepare them. The classification type depends on how difficult it appears that the embalming will be, and that also influences the fee charged. The strength of the embalming fluid increases from Type I to Type V.

I.   Some body heat retained

II. Bodies with moist and clammy tissue, dead from twelve to twenty-four hours

III. Bodies dead more than twenty-four hours, but without advanced decomposition, or bodies that need reinjection of fluids

IV. Certain types of diseases or forms of death, like asphyxia, that require specific types of preparation

V.   Bodies in advanced stages of putrefaction, or with gangrene or ulcerated limbs

There was one instrument used for this procedure that I'd never heard of: the trocar. It's a long, hollow needle attached to a tube, and it has a variety of uses. Generally, it's jabbed into the abdomen, two inches above the navel, and passed up to puncture the transverse colon (to relieve gas), the diaphragm (to get through to the heart), and the heart (to drain the blood). Once it penetrates the heart, it may lunge forward. Apparently it requires a lot of confidence to do it right. However, the trocar is also used simply to aspirate gases from the intestines and insert embalming fluid.

At the end of this internal exploration, any holes that were made are sewn shut or closed with "trocar buttons," which look like plastic screws.

When the procedure is finished, the body is dressed and positioned to look peaceful, and the fingers may be glued together, then stretched for a pose. Generally the undertaker wants to avoid the look of a body lying flat, so one shoulder might be raised above the other, or the coffin may contain a raised bed that tilts slightly. The hair is washed and styled, and the makeup applied. Erica does this herself, and she told me that there's a whole industry out there that supplies cosmetics and clothing for the dead. Although people don't generally see anything below the waist, several companies urge funeral directors to purchase special socks and shoes for the corpse. They also supply suits and dresses that are cut open in back so that they just drape over the body. That frees people from having to struggle to pull clothing onto the corpse, but it appears that the deceased is fully dressed.

When I asked Erica about her own experiences, she related a strange incident that happened during her internship.

"I was dressing a man and I couldn't tie his tie. I always had trouble with that. I got him dressed and I said to him, 'Listen, I just cannot tie your tie. I will try it again, but chances are, it's not going to work.' And, poof! Just like that, I did it. It was so amazing. I tied it perfectly. Later I was listening to his eulogy in the funeral home and as the minister was talking about him, he said how much the man had loved his tie collection. I was so shocked. So I went home and tried to tie a tie on myself. It didn't work. And I still can't do it."

After leaving Erica, I found a specialist in cosmetics for the dead—a "restoration artist"—who was happy to talk off the record. Applying makeup to the dead can be a simple procedure not unlike what many women do every day, but it can also be quite a challenge. Victims of violent homicide, accident, or suicide involving trauma to the face have to be reconstructed, so the makeup artist must take care not to disturb the work. Makeup can also cover the effects of some illnesses that discolor the skin, although putting peach dye into the embalming fluid may work as well.

"Sometimes the person who does the embalming does the makeup, too," she told me, but in her case, it's a separate process.

This cosmetician uses a glamour kit that includes

- an airbrush hose (to create natural-looking wrinkles)

- sandpaper

- cleaners and bleaching agents

- mascara and liner pencils

- base makeup (liquid and cream, for all skin colors)

- lip waxes and cheek color

- hair crayons

- tinted hair spray

- beard clippers and hair pluckers

"My intention," she said, "is to make them look the best they can possibly look, but without seeming overly made up. You can get a suntan effect, if you like, or look like a movie star. You can also just look like your normal, everyday self, which—believe me—is not how you look after the embalming process. Cosmetic makeovers are more important than people realize."

"Why the hair crayons?" I asked.

"That's just for a slight touch. If a woman has only begun to turn gray, you can cover it over and make her look younger. Or if there was blood or some kind of chemical in the hair, washing might not get it all out."

"And the sandpaper?"

"Just for smoothing out scars or places where skin isn't settling correctly. I don't use it often."

She laughed and said that she'd had requests from people getting ready for big events like weddings to do the make-up. "I'm pretty good at it," she said, "but they have to be lying down."

◆ The Freelancer

I learned further details about the art of embalming from Charles Zannino. While he's a licensed mortician and funeral director, he works strictly as a freelance embalmer. What this means is that he spends all of his time embalming and he does that by going on demand to different funeral homes in a defined area. In the course of a twenty-one-year career, he has embalmed thousands of bodies. On call for nearly twenty clients around Maryland, he just hopes they don't all get busy at once.

Zannino grew up in a funeral home with his seven brothers and sisters, and he remembers that his friends disliked coming over because they had to pass the viewing room. Tall and lean, Zannino exhibits a strong Italian manner, frequently punctuating phrases with "Capisci?" His love for his work comes out in a delightful sense of humor.

Although he had run his parents' funeral home for fifteen years, he now prefers to specialize in embalming. His father had done that in the 1940s, charging ten dollars per treatment—a far cry from the hundreds of dollars it can cost these days.

"What I like about working this way," he said, "is the independence. I do work long hours, but I do a thorough job. The beautification aspect is important to me. It's like being a craftsman. I have a talent and this is a trade in which I can exercise it and excel. I spend a lot of time on restoring someone, and I've gotten a reputation. People ask for me." He's even embalmed a few funeral directors whose families requested him.

Zannino feels that people complain about how much a funeral costs mostly because they don't realize all the time that goes into getting all the details right. He puts his work before everything else and may end up working eighteen to twenty hours on any given day.

He went on to explain that there are two kinds of death: cell death and somatic death. In cell death, all the cells die, but somatic death involves three major organs: the heart, brain, and lungs. Cell death happens first and can in fact take place without somatic death, as in cases of gangrene.

The most sensitive and challenging embalming he ever had to face happened when he was in his mid-twenties and had only been practicing about five years. There had been a terrible tragedy in which a woman, depressed over the breakup of her marriage, had stabbed her three sons to death, ranging in age from three to nine. Then she had cut herself with razors and immersed herself in a tub of water.

These four awaited Zannino's attention.

"I was just overwhelmed," he recalled. "You really just feel this tragedy. That was my most difficult case. You work on the children, which is difficult enough, and then you have to work on the person who did this. It was just terrible. The aura of the tragedy hung over the whole process."

He had another case in which a woman had been found after she had been dead for several hours. Her dog had chewed off her chin, jaw, right cheek, and lower lip. This was a true challenge—one that others might have said was too much work—but Zannino tries to approach every case with a positive, can-do attitude. "I had no photo, so I had to visualize in my mind what this woman had looked like before the trauma. I used a special wax to make both sides uniform and I really surprised the family. They didn't expect that we could make her look normal again."

He also had a case where he had to reconstruct a man's beard. "This was a shotgun wound," he said, "so we used the usual waxes and creams for the facial reconstruction, but also a hair swatch to replicate his beard. It worked out nicely." He draws the line on reconstruction only when—because of decay, destruction, or disease—there is no foundation to work with. The tissues that remain in such cases swell too much and discolor to a point that can't be reversed.

The final part of the art of creating a pleasing memory picture for the family lies in restorative techniques: hair and cosmetology. The idea is to get the deceased back to looking normal again, in terms of shape and color. Generally the same person who embalms the body does this, but in some cases an expert is called in.

Restoration may first involve closing wounds. Missing parts are replaced with wax or clay, and suture lines are covered with a chemical. Damaged hands can be covered with gloves, and hair can be dyed. The restorer then applies pigment powder or liquid to the chin, lips, ears, nose, forehead, and cheeks. This helps to accent parts of the face, offset poor lighting, conceal discoloration, and create the image of rest. Cosmetologists rely on photos of the deceased when alive to get the coloration right.

Going from place to place, Zannino has learned a few things from other funeral directors, such as the best ways to make hands appear normal (put tape around the fingers overnight) and how to position bodies for the best visual effect.

Inside his embalmer's kit, you can find the following items:

- special reconstruction waxes

- cosmetics and creams

- injector and suturing needles

- tacks for securing the mouth and jaw

- eye caps to slip under the lids

- canula tubes for transferring embalming fluid

- gloves, goggles, and surgical mask

- tissue scissors, scalpels, and forceps

"Does the work ever get to you?" I queried.

He admitted it sometimes did.

"In a dream," he said, "I'll be in a viewing room and I'll see bodies moving in the coffin, so I try to get them to settle down."

After listening to him, I decided to do more research and found the following odd tales.

Strange Embalmings

A real embalming thriller occurred in 1924 when Russian leader Vladimir Lenin died. Determined to preserve his body to elicit religious sentiment from the masses, the Soviet officials debated over the best methods. Before long the corpse showed signs of deterioration: the lips had begun to part and the skin to discolor. A committee frantically sought a way to prevent this, so they refrigerated him. However, freezing can accelerate the process, which they soon discovered. Then they injected the body with chemicals, but they leaked out and appeared to do little actual preservation. After a month passed, the committee called in a noted pathologist who found the body in poor shape. Since he'd had success with organs immersed in balsamic liquid, he suggested they try the same with Lenin's body. They rejected his ideas, but time was running out.

Those men who did step forward to undertake the procedure knew that their own lives were at stake: If they failed, they could die. However, if they succeeded, they would acquire fame and significant political privileges.

They developed a unique technique, known by only four people in the world at the time, which had to be performed every eighteen months. It involved immersing the corpse in a glass tub filled with a secret combination of gylcerol, potassium acetate, and a few other chemicals. Then they bound the body in rubber bandages and dressed it. Afterward, and every two weeks, they wiped embalming fluid on the face and hands, and to assist in preservation they stored the body at sixty-one degrees. Lenin's corpse is still in storage to this day.

Even more bizarre was the embalmer who took over the task of making the corpse of Evita Perón—Argentina's former first lady—exist as an eternally perfect specimen. She was something of a saint to the masses, so when she died in 1952 at the age of thirty-three, Juan Perón employed anatomy professor Dr. Pedro Ara to preserve her. Ara was a skilled practitioner of "the art of the dead," bragging that he could ensure that Evita never decomposed. She lay in state for two weeks and then Ara took over. He injected and reinjected the body with chemicals, and then submerged it in baths of acetate and potassium nitrate. Finally he coated her with layers of plastic and wax until she was perfectly preserved. The entire process took about a year, and when all was said and done, it cost Perón over one hundred thousand dollars. Then Perón was overthrown and Evita's body was transported all over the place in some clandestine effort to get rid of it. For a while, she was even stored in apartments and attics, and finally she was secretly buried in Italy. But that wasn't the end of things for her.

When the exiled Perón settled in Madrid with his third wife, the Argentinian military requested his help with a chaotic country. He demanded Evita's return, so they dug her up and sent her to him. She still looked pretty good, with only a few cracks in the plastic, so Perón had her cleaned up and then he placed her in the dining room where he and his wife ate dinner.

In the Business

John McDonough is a licensed funeral director in Lowell, Massachusetts. On Sunday nights he runs a chat session for the Electronic Funeral Service Association, where they talk about everything from embalming products to body-lift technology. John is a middle-aged man of average height and build, with a friendly face and professional demeanor.

"You come from a family of morticians?" I asked him.

"Yes, five generations," he said, "but in this part of the country, it's 'funeral director,' not 'mortician.' A mortuary is a funeral home that has a crematory on campus, and a funeral home does not." He also explained to me that corporations were buying up many independently owned funeral homes, so that the family-business aspect of funeral homes was beginning to die out. However, those who honored that tradition were hanging on.

"So how did you get involved?"

"I was thirteen when I first began to work here and I loved it. They let me start the car and move it, which was cool. Eventually they asked for my help on funerals. A typical wake has twenty or thirty pieces of flowers in the room, and on busy days, they'd have me bring the flowers in. My mother didn't think it was wise to actually let me see the dead person, so they used a curtain to block it."

"So then how old were you when you actually saw your first body?"

"Thirteen," he said with a laugh. "I used to peek."

What did he like about this business? I wondered.

"I like the fact that I can make people feel better about their situation." Then he told me a story. "Very early on, I spent most of one Saturday bringing flowers out to this guy's wake. When the widow came in, she saw me and apparently said something to my grandfather. He came into the back office where we'd hang out and said, 'Mrs. So-and-so wants to meet you.' So I went up the hallway with him and she asked him if she could have a couple of minutes just with me. My grandfather said, 'Sure.' So she said, 'Sit down, I want to talk to you.' She pointed to her dead husband. 'You see that guy in the casket right there?' I said, 'Yes, ma'am,' and she then said, 'He has over a million dollars in the bank. We worked and worked and saved and saved. He retired last month so we could enjoy this. Now I can go on all these world cruises, but I go without my best friend. I want you to remember this: Enjoy every day as you get it.'"

I wanted to know if he'd had any quirky experiences as a funeral director and he told me the following incident.

"We've had people put stuff in the casket. For example, one man really loved golf, so they put a golf ball into the casket, but didn't tell us. So we closed the casket and started taking it up the church steps for the service. As soon as it tilted, we heard this boink, boink, boink, and we were wondering, What the hell is that?"

That surprised me, but it made sense that people would be buried with things they had loved.

"What other things do people put in?" I asked.

"With one guy, his kids put a cell phone in his pocket and it rang at the cemetery."

I laughed and asked, "Did he answer it?"

McDonough thought about other requests and said, "We have one guy doing a pre-need package who is interested in having his cremated remains scattered on the moon. There's actually a company that does that. They rent space from NASA on the probe that goes to the moon. If we can manage it, we'll do it."

Other Options

Not all arrangements go through funeral directors, and most people who die in the U.S. are buried rather than cremated. That means that other people are offering similar services, and it wasn't long before I met the owner of a private cemetery. That was the first time I realized that people could actually manage burial plots as a business.

These days in the U.S., the roughly one hundred thousand

burial grounds in the country are divided among municipal, religious, veteran/military, and private cemeteries—some of which are known as memorial parks. Except in those few states where the practice is prohibited, increasingly more acreage set aside for burial is getting bought up by for-profit concerns. In any of these places, you can purchase a grave site, a crypt, or a mausoleum. If cremated, you can reserve a niche in a columbarium (a collection of urns).

Ted Martin owns a medium-size memorial park in Columbia, Pennsylvania, which sees more than one hundred funerals each year. He himself can perform many of the same services that a funeral director offers, such as acquiring the death certificate, transporting the body, and opening and closing a grave. Since many people are afraid of cemeteries, Martin tries to make his burial grounds inviting. The park sits up high, for example, so he allows an area church to hold sunrise services there. "We're here for the living," he said. "We're part of the community's legacy."

It's mostly a quiet place, but he's heard some stories. His veterans' section, in fact, very nearly acquired the reputation of being haunted. One night, two kids were parked there, making out. They saw something that so alarmed them that they headed immediately to the local police station. Pale and shaken, they swore they'd seen a ghost. It was out there in the cemetery hanging from a tree. Whatever it was, it shook the branches at them and acted aggressively. They had feared that it might harm them and they insisted that someone investigate. The two were so convincing that an officer finally agreed to drive out to have a look. As he went toward the veterans' section, he spotted something moving. He'd expected this to be some prank, but just as the kids had described, he could see that there was something as big as a man over by a tree and it appeared to be hanging. A little nervous, the officer drove closer and shown a light on the entity. To his amazement, it was a full-grown buck with its antlers stuck in the branches. The officer called for help and they got him down, and that was the end of the ghost story.

Martin told me another tale as well, which was just as amusing. It seems that a man who was very possessive of his money was afraid of what his wife might do with his fortune should he die before she did. So he had a will drawn up in which he required that his money be buried with him. He insisted that his wife sign the document, and to his surprise, she did. Eventually he got sick and died, and the wife came to the viewing. The lawyer who'd made up the will noticed that there was no money in the casket, so he asked her about it. She smiled, drew a check out of her purse, and placed it into her deceased husband's jacket pocket.

"There's your money, dear," she said. "You can cash that at the bank."

Counselors

I arrived at the Washtenong Memorial Park in Ann Arbor, Michigan, and a tall man dressed in a gray suit approached me. He wore glasses and had short blond hair. "Can I help you with anything?" he asked.

I saw from the pin on his lapel that he was William Root, a "family service counselor." Under the impression that he was some kind of grief therapist, I asked him about his job. It wasn't until he explained his background in sales that it dawned on me what his position entailed: he was a cemetery salesman. I'd been fooled, but I guess that was the point. Believing that he was a counselor would be more inviting during one's time of grief than being approached by a salesman. Added to Root's market-

ing experience was a connection to the industry: His grandfather in Iowa had run a company selling redwood coffins with
copper nails. As a boy, Root had romped through the display
room, so he was quite comfortable showing people the casket
selection.

Besides assisting people with their funeral needs, he super-
vised services in the mausoleum and sold "pre-need" packages
to those who want to get their affairs settled while still alive.
He admitted that he was in direct competition with funeral directors because, like Ted Martin, he offers the same basic services.

Root then explained that Washtenong is a sixty-six acre "park"
that offers in-ground burial, mausoleum vaults, cremain niches,
and even a "scattering garden." They also had a memorial to
those who had donated their bodies to science. This place was
different than many cemeteries that I'd seen in the area, notably
in the absence of statuary, and he explained that memorial parks
require that grave markers be flush to the ground to create a
more pastoral appearance. They were quite popular around the
country in the 1920s and 1930s, and thought to be the cemetery of the future. I thought it was strange that they marketed
the burial sites like real estate—buy your "private mausoleum
estate" in a "park"—but when you're pushing a product like
graves, you need to make it inviting. They had plenty of space,
Root assured me, for some time to come. I guess he was hinting
that, if I was so inclined, they could accommodate me.

Although he had been there for only a short time, he'd had
one unusual experience. A man who owned space in the cemetery had been living in Florida with his wife. She had died and
he wanted to bring her back to Michigan. "He had her sealed
into a coffin," Root said, "and he put the coffin into the back of
his pickup and drove the whole way up with her there in the

back. He apparently decided, since he was going to Michigan, to have a vacation as well, because when he drove into the cemetery, he was carrying the coffin and towing a boat."

Monument Dealers

Funeral directors, counselors, and cemetery owners can all sell monuments as well, but I learned that there were people who also made their living focusing exclusively on that. It was one more area of cutthroat competition.

Richard Groom lives in Lancaster, Pennsylvania. His father had owned a burial-vault company, so he was raised in the business. He'd also had an uncle who had operated both a morgue and a funeral home, and for lack of space, the man had put the morgue into the first floor of his own house. Groom recalled routinely bumping into bodies on his way up to his uncle's apartment.

Groom himself has operated several different cemeteries, and currently he owns a business that sells laser-etched monuments. After news of a passing is published in the paper, he sends out a letter to the family with a guarantee that he will sell them a monument that is 25 percent below whatever price they get elsewhere, or it's free. "And we've never given one away," he assured me. He said that most people procrastinate making a choice until about six weeks after the death.

He pointed out that cemetery owners try to discourage people from buying from independent dealers like him, because it means money out of their pocket. Some of them resort to adding "nuisance fees" to the cemetery bill for "use of their roads" or they require an abundance of paperwork to hinder the installation. Groom says it just makes things more difficult for the family.

He had no tales about monuments, but from his cemetery-management days, Groom recalled an incident with a disinterment. The family had insisted on attending the operation, but Groom knew too well that things can happen that families ought not to witness. "The last thing you want, with a family member standing there, is the body rolling out."

To avoid problems, he went out two hours earlier to start the work. As they lifted the vault, the suction broke on the bottom and the body came out. "When you pull a three-by-eight-foot piece of concrete out of a hole, you may pull the bottom out of the vault. We took it up in the air and the casket was too heavy, so it fell out and broke into pieces. Even the body came apart." In haste, he got another vault, put the body back in, sunk it, shoveled some dirt on it and made it look ready for the family. To his relief, they never knew a thing.

Funerals

The National Museum of Funeral History is located in Houston, Texas. It houses rare artifacts and provides historical information about the American cultural rituals surrounding death. It also has a display about the funerals of political figures and celebrities, including Elvis, Martin Luther King, and Mickey Mantle. You can see a replica there of Abraham Lincoln's coffin (and a photo of him in it), along with an array of hearses, from the 1832 horse-drawn vehicle to a 1916 "funeral bus." Also on display are mourning clothes, embalming instruments, jewelry made from the hair of the deceased, and many different types of coffins—including the solid glass casket marketed in the 1800s. The area devoted to the history of embalming includes a re-creation of Dr. Thomas Holmes embalming a soldier during the Civil War, and one of the most disturbing displays is a casket built for three—intended for a couple who wanted to kill themselves when their child died of a fatal illness.

There's also a mortuary museum in Heber Springs, Arkansas, which is just the front room of a garage for car detailing. It's open whenever Warren Olmstead feels like opening it, and its central feature is an 1896 horse-drawn hearse. He's also got portable Victrolas that had been used at the graveside, a variety of coffins, and antique embalming tools.

Funerals are for the living, a way to announce that a death has occurred and to use ritual to work through grief. Whether an open-casket viewing or a simple memorial service with cremated ashes and a photo, it's a time to gather relatives and friends to honor the person's passing. Some people need to actually see the dead body to get closure, and funeral directors offer restorative services for that purpose. Others create their own traditions.

While people say that funerals are pretty much all alike, I know that some break all the rules—socially, that is. One of the most unusual funerals I ever attended was for my friend Corey, down in New Orleans. As dictated in his will, he was cremated, but there was no urn, no niche in a mausoleum, no cemetery plot or tombstone. Just a box of cremains. He had made provision for his closest friends to dine out on him—and he ended up being the centerpiece. At a local steak house, we placed his box on a stand in the center of the table and told the waiter to get him anything he liked.

It's also true that different religions have different rituals. I'd already heard about the Gypsies having a pig roast and several other religions throwing money into the grave. Muslims believe that the sooner a good man is buried, the sooner he reaches heaven, and he's always buried facing Mecca. Some Buddhists place the corpse in a prayer position for three days before burial to allow the soul to separate.

It was through a friend, Dot Fiedel, that I learned about Jewish funerals. Not Jewish herself, she had to learn the rituals in order to bury her husband, Sam. Generally Jews are buried in a shroud (unless wounded, in which case they are buried in the clothing on which they bled because they must have all of their parts with them).

"The rabbi 'prayed' Sam's coffin to the grave," Dot recalled. "The pallbearers took several steps, and then they would stop and the rabbi would pray. It was a slow progression to the final resting place." The reason they stop, I learned, is to give the mourners moments along the way to reflect on the person.

After the burial, there's a waiting period before setting up the grave marker, which is installed before the year is out with an unveiling ceremony. (Orthodox and Reform Jews diverge on some of these practices.)

"Sam had requested that I cover all the mirrors in the house with black cloth right after he died," Dot said, "and I was to keep them covered for seven days—the amount of time for shiva, the prayers said for the dead. Sam explained that if he was still there in the house after he died, and he found the mirrors covered, he would then know he was dead."

"Did he ask you to do anything else?"

"He also asked me to go to his grave with a bottle of cognac, drink it, and dance on his grave. Thirty days after his burial, my family and I were to walk around our house, counterclockwise, three times. Then we were to face the eastern sky and wave our hands and call goodbye to Sam out loud for several minutes. It was something he needed to hear to help him know he was now among the dead."

◆ *A Gathering of Undertakers*

The 119th annual meeting of the National Funeral Directors Association was held in Baltimore, Maryland, in 2000. It's the world's oldest and largest international funeral-service-trade show. They estimated that there would be between six and seven thousand participants, including a record number of vendors who would put up the largest exhibition of death-related merchandise in funeral history—to the tune of one hundred thousand square feet. Some fourteen thousand licensed funeral directors and embalmers belong to the NFDA, and the convention provides a place to network and see what's new in technological advances. The only way I could go, however, was to find a sponsor, and through serendipity, she appeared.

Deborah Brown is a slim Cajun woman with lively dark eyes and black hair. She had written a novel that involved funeral directors, so while we were talking about it, I learned that she was a consultant on standards of care for funeral homes around the country. In business for a decade, she had over thirty clients—that is, she regularly visited thirty funeral homes.

Deborah confided that she had always wanted to experience what it was like to be closed inside a coffin, so she'd gotten a friend in the business to assist her with this "ultimate fantasy." She took along a cassette recorder to record her thoughts, and the first thing she noticed was how different things sound from inside.

"Your own heartbeat starts sounding like someone whispering things to you," she told me. She also realized how difficult it was to fight off the feeling of being closed in. Fifteen minutes "buried alive" was sufficient for her. Being utterly claustrophobic, I don't think I could have even gotten in at all.

However, thanks to Deborah, I did get into the conference.

As I drove to Baltimore that weekend, I knew that I was close to the NFDA gathering when a car passed me with a fake leg and arm hanging out of the closed trunk. That had to be an undertaker with a morbid sense of humor.

The first event was a party thrown by John McDonough for his chat group, the EFSA. Right away partygoer Marilyn Gubbiotti told me a story about a man whose death had put a serious financial strain on his family. They weren't quite sure what to do. Then, out of the blue, Marilyn received a call from a woman from out of town who had heard about the man's passing. She was calling because she wanted to pay the entire bill herself. "In my experience," said Marilyn, "that was a unique gesture."

Everyone there seemed quite jovial. No morbid Vincent Price, no Peter Lorre, no Boris Karloff. In fact, they laughed easily, greeted each other heartily, and displayed a warmth and friendliness that belies the stereotype. Because they all had seen the many faces of death, worked long hours, and knew the strain on family life, they were like family to one another. "This is an industry that puts the dead first," someone pointed out, "and that takes its toll."

As I listened to them talk, I heard an amusing story that also made me wonder how much the dead may linger. The story went like this: A funeral director was leading a procession into a large city when he came across a construction detour. Forced to take another route, he ended up getting lost. With "Uncle Al" in the hearse, he led the entire line of cars through a shopping center and into several neighborhoods before he finally found his way to the cemetery. When he arrived, he prepared himself to offer profuse apologies. However, instead of the anger he expected to face, he found the surviving relatives laughing hard as they got out of their cars. He asked what was so funny.

"It's just like Al," they told him. "In life, he was always getting lost."

The next day, I went to the convention center, where thousands of people in various aspects of the funeral trade were gathering. I went up the escalator to the meeting rooms and directly in front of me was the NFDA's booth for collecting funds for a World War II memorial. That reminded me of a chilling tale that a woman had told me the evening before.

"The hardest pre-need arrangement I ever did," she said, "was for a man who had served in both World War II and the Korean War, and he'd seen a lot of combat. His wife had died a few months earlier after some sixty years of marriage. He was heartbroken at that time, but not ready to die. Then he came in one morning and brought with him his doctor's report, which had been written out for him because he was deaf. The report said that he had only a few more weeks to live, and it would be a difficult death. As we read this, he cried and said that he needed to make his arrangements. He told me what he wanted and I wrote it and let him read it. By this time, I was crying, too. He seemed so alone and so frightened. Trying to bridge the gap between us in age and communication abilities, I wrote down, 'Are you afraid to die?'

"He looked at me and said, 'Yes. I've killed a lot of people in my time and I'm afraid that they're over there waiting for me.'

"It was a stunning moment. I didn't know what to say. He'd lost everyone and now he had to face alone his greatest fear."

From the memorial, I went to listen to the first speaker, and then started meeting funeral directors. Joe Ambrose, a local professional whose son had joined him in the business, greeted people with an elfish smile. Everyone seemed to know him. As we sat together outside one seminar, he pointed through a window to a building across the street and told me about a man who had died there. That was one of his own removals. He went to collect the body and then notified the man's wife. However, someone else called in also claiming to be the man's immediate family and having no notion of who this "wife" was. It turned out that the deceased was a bigamist with two separate families that, until that moment, he had managed to keep secret from each other. The situation was awkward, Joe said, but they ended up resolving it peacefully. I guess it pays to have time before you die to get your "affairs" in order.

As we walked through the halls, another funeral director said that he'd heard about a woman who was a diehard baseball fan and had paid to have her ashes buried beneath third base in the Brooks Robinson Memorial Stadium.

It was "unofficial," he said, but she was there.

Then funeral director Daniel Hartzler related a tale that he'd found amusing: "This hearse salesman was down on the Eastern Shore. He'd just delivered a new hearse and had an old hearse that he was returning. He fell asleep at the wheel, drove off the road, and had an accident over the bank. Although he'd broken his arm, he managed to crawl up the bank, but then he passed out. A woman found him and called the state trooper. Just then the driver came to and he overheard the confused woman ask the trooper, 'Where could he have come from?'

"The trooper pointed down the embankment to the hearse, which was illuminated by the dome light.

"Her eyes widened and she quickly asked, 'Did he come out of the front or the back?'"

The stories were nonstop about quirky incidents and the practical jokes that some funeral directors played on each other. "We do it to let off steam," one man told me. "The pressure of this job is enormous and many families fall apart. You'll find multiple marriages among us and lots of professionals who have risky hobbies in whatever spare time they have, because they see so much and it's difficult to talk about."

"In fact," said another, "we don't talk about it because it's hard to find words for all the things that people do to themselves or what can happen to them. What can you say about an accident scene where an entire family on their way to see Grandma has been killed by a drunk driver? You pick up children and see their toys. You listen to a grieving relative who's just lost everything. And you have to keep all of that to yourself. Often, even your spouse just doesn't want to hear about it."

Yet despite the dark undertones, it became clear that one of the allures of the business is the fact that no one ever quite knows what to expect. Each day has the potential for something unique to happen.

One of my favorite stories was about a funeral service for an older man who was a member of a nudist community.

"We didn't realize quite how unusual this would be," I was told, "until his family and friends came in to talk about arrangements for the visitation and service. We asked, as we normally do, how they wanted the deceased to be dressed, and they said, 'Well, this is the way we feel about it. He did not wear clothes when he was living and there's no reason for him to wear clothes when he's not living. Would that be okay with you?'

"It was different, but we saw no reason not to do as they wished since it was to be a private service. The only stipulation we made was that we could not have a graveside service for people without clothes.

"So after the embalming, we left off the man's clothes and had an open casket with the deceased's nude body. It was a little awkward, because often the clothes are used to disguise things on the body that people don't want to see. In this case, we didn't have that advantage. The embalmer did try to figure out a way to make the man's arms stretch far enough that his hands might come together and cover his private parts, but he couldn't quite make that work. Still, we pulled the drape on the casket up a little, and that helped *us* to feel more comfortable.

"Then there was the service to deal with. We had to do a few things differently than we normally do. For example, since our chairs have fabric, we had to rent chairs for unclothed people to sit on, and most of those who came wore nothing (although some of them perhaps ought to have given that more thought). We generally offer picture boards to pin up photos of the deceased person during his or her life. This man's relatives and friends took advantage of that, and except for when he was growing up, almost all of the pictures were of him nude. His visitors were all comfortable with this display, although I can't say that we were. In fact, the funeral directors were the only people in attendance who were wearing clothes, and as the room filled up with naked people, they began to feel overdressed. Afterward, they agreed that this was the most unusual service they had ever conducted and they certainly hoped they would not be called on to do another one."

What I most wanted to do at the convention was tour the exhibits, so during the afternoon, Deborah Brown accompanied me. This was a first for me and I can honestly say that I've never seen anything like it. There were endless displays of caskets, urns, body bags, state-of-the-art limousines, body lifts, transport boxes, pet caskets, cemetery memorials, and online marketing. I saw one booth that offered DNA analysis, and another had all kinds of chapel furnishings. If I'd been in the funeral business, this would have been the ultimate candy store, and one reason there was so much variety was due to recent attempts by the industry to cater to the Baby Boomers.

In recent decades, the American funeral business has come under attack, in part because certain unethical practitioners have seen it as a way to take advantage of grieving and vulnerable people. For example, for a while it was the practice to package all of the procedures and services together, giving people no choices and thereby getting them to pay for things they might not have wanted. Some undertakers told people that embalming was required by law and urged them by trickery to purchase the most expensive caskets. Thus, the government stepped in and the Federal Trade Commission issued the Funeral Rule, which forced funeral expenses to be itemized accurately up front so consumers would know what they were getting. Even so, some undertakers continue to use deceptive practices for big-ticket items like coffins. They also urge people to accept services that are unnecessary, knowing that people deep in grief are unlikely to check the laws.

However, the stories that get told give the funeral industry as a whole this negative veneer, which is unfortunate, since many professionals are trying to do a job that benefits others. Even so, there's no doubt that it's a business, as I saw all around me in the exhibits, and that people prosper from sales.

Every year, Americans purchase around two million funerals, distributed among about twenty-seven thousand providers, and that number will grow exponentially with the passing of the Boomers. While cremation is more popular than ever, it still accounts for only about 30 percent of all funerals.

The industry seems quite aware that consumers' desires are changing. Much is made of how the Baby Boomers seek something more vibrant and personal than past traditions allow. As one funeral director said, "Our business is show business and hospitality. We have to choreograph the experience."

As I walked through the casket displays put on by Batesville and Aurora, I was surprised by the variety. It's no longer a matter of just choosing one type of wood over another or picking the color of the satin lining. One can get elaborate paintings on the outside of the casket now, such as a patriotic war scene, the ocean, or the woods. The urns, too, might be in the form of a golf bag, fishing tackle box, leatherbound book, or favorite bird. I even saw one that came with wind chimes.

It felt so odd to me to be walking around a casket display the way I might stroll through an antique car show. While I was looking at a massive oak casket with an interior painted panel of *The Last Supper,* Deborah came up beside me.

"You must see a lot," I said. "What's your most memorable experience in a funeral home?"

She thought about that and said, "Probably the experience that stands out most to me involves something awful."

"What do you mean?"

She's seen burn victims, shotgun victims, and suicides, she told me, but the encounter that truly disturbed her involved an embalming.

"I was there for a routine evaluation," she said, "a process that would help me establish consistent standards for operational procedures. Things began simply enough as the embalmer— we'll call him Max—washed down the body of a petite, ninety-year-old woman. He had nearly two decades of experience, so I expected competence, but to my surprise his hands actually shook as he made the incision to locate the carotid artery. Then what should have been a two-inch incision wound up being a large gash that would need multiple stitches to close. I was surprised. Thinking that my presence might be unnerving him, I backed away. Within minutes, he relaxed, and his hands worked deftly as the embalming fluid ran its course. I relaxed, happy that someone's grandmother was getting good treatment, until it became clear that Max was restructuring the routine procedures to make things more convenient for himself."

"In what way?"

"Suddenly, the woman's neck and face began to swell to five times its size and her skin took on an orange hue. I bit my tongue, watching as Max cursed and tried correcting his error, all to no avail. Shrugging, he moved on to the woman's toothless mouth. Instead of using the standard injection system to close the orifice, he ran thick string through the sinus cavity and jaw. This made the woman's mouth and head jerk about like that of a morbid puppet.

" 'It's faster this way,' he explained to me. 'They're dead, so they don't know.'

"Then he continued to treat her as if she were just a side of beef rather than a person. He completed the job by aspirating the body with a dull-edged trocar. He jabbed the long, hollow metal tube against the side of the abdomen again and again until it punctured through with a sickening *schhhlop!* He grinned in triumph and then shoved the trocar through the internal organs, puncturing and suctioning vehemently. I watched in horror as the pointed end of the instrument poked against the opposite wall of the abdomen, threatening to skewer the woman.

"Just about that time, the funeral-home owner walked in, looked at the body, smiled, and asked if I was impressed with their 'top' guy. I glanced back at the deceased, whose upper torso now looked like that of a two-hundred-pound liver-disease victim, and gulped. I'd learned a lesson that day. No one truly knows what goes on in the back room of a funeral home, and embalmers who say, 'This is the best I can do under the circumstances,' may just be lazy or indifferent."

I grimaced. I had to admit it was pretty disgusting. "How did it affect you?" I asked.

"Whenever a loved one dies now, I feel compelled to be present during the embalming. Although ninety percent of the funeral homes around the country truly do wonderful work, you never know when a 'Max' will be on the job."

As we talked, we looked at the displays and came across a booth that was set up with colorful and vivid abstract works of art. Puzzled, I inquired further and discovered that Jane*Us Inc., "Eternally Yours," is a unique type of memorial vendor.

Bettye Wilson-Brokl runs the company, and she said she came up with the idea after her mother passed away and asked to be buried in a family plot quite far from where she lived. Bettye felt the need to keep her mom close. Her solution was to paint a picture and use some of her mother's ashes in the painting itself. Then she made the same artwork for other members of her family as well. "There is not one day that goes by," she says, "that I do not reach down and touch my mom."

Her company offers three sizes, and some include "openings" for another family member (or more) after they die. The family can provide a picture, select different colors, or just choose from the company's selection.

I asked Bettye if she had gotten any unusual requests and she recalled one from a musician.

"We did a pre-need request from an entertainer in New Orleans who wants a sheet of music for the background, specifically his opening song. We've prepared it, and when his time comes, we'll incorporate his cremains as the musical notes."

Given how eccentric people can be about death, I wondered if she'd had any amusing incidents. Bettye told me that she'd heard one story from an executive of a funeral-merchandise corporation who sold her product. "He told me that an elderly man sitting in his office asked about the memorial art hanging on the wall. After explaining what it was, the executive asked his thoughts about it. The elderly man came up with one reason after another why he should *not* consider memorial art for himself. Finally in pure frustration, he blurted out, *'What if I end up in a yard sale?'* "

I came away from the conference with a new appreciation for what funeral directors do, but also mild disappointment that I hadn't found Vincent Price dressed in black, rubbing his hands with a creepy smile. What I came to understand is that this is a business in which making a sale is important, but there must also be a balance of genuine concern for the clients.

Learning the Trade

There are fifty-two mortuary schools in the country, with around twenty-five hundred new students entering each year. Many are from family businesses, but increasing numbers of students have had little exposure to the trade; they see it as a good career move.

One guy, Ed, was going out on his first removal duty. He was with a partner, but was understandably nervous. The body was in the cellar of a tract home and the police had determined that the fortyish woman had died of a heart attack while feeding her animals . . . and she proved to have had quite the menagerie. They had made a superficial count of around thirty-seven animals, from cats to iguanas, and including a mice-breeding area for feeding a large snake. Ed had a terrible fear of rodents, but the cops assured him that everything was caged. Uncertain whether to trust them, he and his partner entered the house.

The place was cold, so many of the animals were curled up in their cages. Ed walked quietly by and tried not to disturb them. In the cellar, the light was dim, but he had no trouble finding the corpse. She lay on her back, her eyes wide open, and she was obviously dead. He looked around to be sure that none of her creatures were poised to attack, and then bent down to position himself at her shoulders to lift her onto the gurney. His hand touched something that wiggled and he immediately backed off.

"What's wrong?" asked his partner.

"There's something under her."

The partner laughed and came over to where Ed had knelt beside the body. He took the corpse by the shoulders and lifted her. Beneath her was a cluster of mice, obviously seeking what little warmth the body had left. When the light hit them they ran, and so did Ed, right up the steps and out the door.

In Anaheim, California, two mortuary students actually fell in love and got married . . . on Halloween in a cemetery. Surrounded by eighteen hundred crypts in the Melrose Abbey Mausoleum (a few with "residents" on which they'd worked), and with many deceased souls in graves outside, they joined their lives together as one. The groom wore an old-fashioned undertaker's suit, and the bride was assisted by a cosmetologist who does makeup on the dead. Their wedding cake was topped with a Mexican "Day of the Dead" bride and groom (little skeleton figures dressed for a wedding). Since their lives are so consumed by the dead, it seemed to them appropriate to wed in the presence of former clients. However, they also invited guests who were still standing.

Another school incident was related by Father Massey, once an undertaker in Hollywood and now a bishop in Arizona. While in college, he got a job in a funeral home to pay expenses, and when the owner offered him a full-time job if he went to mortuary school, he accepted.

This was at a Jewish funeral home in Hollywood, and since he was not Jewish, he ended up working on Saturdays. One Saturday in particular stood out.

"The front doorbell rang," he recalled, "and this little lady came in who looked like she'd been sleeping in the gutter. She told me that her mother had just died and she needed to make funeral arrangements. I approach everyone with compassion, to help them get through, so despite her apparent circumstances, I went ahead and made the arrangements.

"Then it came time to go into the casket room and I explained the basic differences between caskets. At that time, we'd take them in, give them an explanation, and leave them alone to make their choice. We didn't pressure anyone.

"About ten minutes later, this lady came out and she had the price tag for an African plank mahogany orthodox casket. I was surprised. This was in the sixties and that casket was about six or seven thousand dollars. Although I had doubts, I sat down and figured up a bill. She picked a rabbi to officiate, and I was thinking in the back of my mind that this man charged a rather high price. The norm was about thirty-five dollars, and his fee was five hundred dollars.

"When I told her this, she didn't bat an eye. 'That's fine,' she said. She then added that she would need eight or ten escorts for the funeral, and I started thinking that this woman had real delusions. This funeral was going to be far too expensive for her, and how was she going to pay?

"Although I thought she was a crackpot, I went ahead and totaled up the bill. Then I asked her, 'How did you want to take care of this?'

" 'I'll write you a check.'

" 'Okay,' I said slowly. I watched with some reservation as she wrote me a check. Then she looked at me and said, 'While you're upstairs verifying this check, may I use your phone?'

"I looked at her, unsure what to say, and she explained, 'I want to call my driver.'

"I then looked down at the check and saw that it was signed, 'Shelley Winters.'

"She smiled and explained that she had worn this particular outfit to the hospital because her mother had been in and out of a coma for several days. 'She hasn't recognized me. This was a costume that I'd worn in her favorite movie and I was hoping that she would recognize me this way. But when I came into the hospital lobby, the administrator met me to tell me that my mother had died, so I went walking down Hollywood Boulevard and this was the first Jewish funeral home I saw. So I came in.'

"I did the funeral on Sunday and she gave me a nice gratuity, so I decided to use it toward my tuition. On Monday, there was a call for me over the loudspeaker to go to the dean's office. I figured that was about my tuition, and I was pleased because I had the money. I got in there and the dean mentioned the service I'd done for Ms. Winters's mother. I acknowledged that I had, and he said, 'Well, she just left here, and she's paid your tuition for the year.' "

◆ Unusual Requests

Father Massey went on from there to spend thirty-two years in the funeral business, in three different states.

"I had a family come in one time," he recalled, "who wanted to move Grandpa. He'd been buried several years before but they'd bought a family plot, so they wanted to get him exhumed. We made the arrangements. He hadn't been embalmed, so all that was left of him was a skeleton. We brought that up and put the pieces into a redwood box. Then the rabbi came to say prayers for the reinterment.

"Just then I realized that the workers had put the casket over the grave backwards. The head was in the place where the feet were supposed to be, and that's how he'd get buried. I stopped the rabbi and told him I would have to get the workers back to get it rearranged. He looked down at the casket and said, 'Don't bother to call them.' He reached down and picked up one end. I heard this klunk, klunk, klunk, as the skull rolled to the other end. Then he said, 'Let's proceed.' "

Father Massey obviously loved telling these stories, so I asked him for more. He told me one about an actress that was supposed to remain a secret.

"She called me one night to set up an appointment," he said, declining to name her. "I went to her home and she told me she had buried two or three husbands, but her problem was that she wanted to be buried next to her cat, not any of her husbands.

" 'Where's your cat buried?' I asked her.

" 'In a pet cemetery.' She told me she wanted to be cremated after she died and have her ashes placed in the grave with her pet. I checked and there were no regulations against it, so that's exactly what we did. If you can find it, there's a stone, and her name is on it."

Since he was in Arizona, I asked about Native American practices. He laughed and assured me that he had a tale to tell about one experience on the reservation.

"I had the Indian contract one time in Phoenix. Our job was to do the embalming, but then take the deceased to the reservation for specific rituals. One day we had a call for a man who had died in his mid-thirties. We picked him up, embalmed him, put him in a suit and casket, and returned him to the reservation. We came back the next morning to take him to the church, and one of the elders called us over and said, 'We have a problem.'

" 'What's the matter?' I asked.

" 'Well, we took him out of the casket.'

"I couldn't understand why that was a problem until he took us over to show us. There was the dead man, clothed in his tribal garments, lying on the cement next to the coffin. The elder explained that they had taken him out to re-dress him and could not get him back in. When I still didn't understand, they demonstrated for me. Now, we had put him in a casket with a hinge panel that opens up as a small opening on the top. That's so you can look inside and have a viewing of the face and shoulders. They had assumed that this panel was the way to open the coffin. They'd managed to get him out that way, but couldn't put him back in. He just didn't fit. I walked over to the casket to show them how the entire top opened up, and when I lifted it, the elder nearly had a stroke."

A mortician in New Mexico, Robin Wall, was once asked to do something prior to a burial that utterly puzzled him. A woman was making arrangements for her husband, who had been killed in a drilling-rig accident, and she had a strange request. "She had heard that many funeral homes split the clothes so they can just drape them on. We don't do that here, but in this case she asked us to. We went to the service, which was open casket, and at the end of the service after everyone had walked out, she went up to the casket and said, 'Okay, now roll him over.' We were surprised, but she insisted, 'I want him rolled over.' So we did. Much to my distress, he was buried facedown, with his backside exposed. I figured it was her last gesture of 'Screw you.' "

Returning to the East Coast, I heard from Dan Poppoff, who had grown up in Erie, Pennsylvania. He knew from the age of five that he wanted to be a funeral director, and he has also had some interesting requests.

"The most unusual request I received," he told me, "was that of a middle-aged woman who we just made funeral arrangements with, for her husband. He'd died early that day, and she wanted to hold him one more time. That's not so unusual, but she asked if she could do it before we dressed him."

◆ Entrepreneurs

With the death rate lower than expected, and prices increasing, many people are looking around for alternatives to the traditional funeral. Savvy entrepreneurs have stepped in to offer their wares, and other people are finding ways to cash in on some unique death-related niche. A few get involved just for fun. These endeavors include:

Mummification: It runs you around sixty-five thousand dollars (much less than Evita), and there aren't many places, aside from Summum in Salt Lake City, that do it. After you die, you spend about six months in a vat of "secret" preservative. Then, before being wrapped in gauze, you get a lanolin treatment. But that's not the end. Then you're rubberized, bandaged again, and placed in a bronze "mummiform." The cost is in the packaging, and if you've got the money, you can even get it done for your dog.

Death Investigations: In Los Angeles, you can call 1-800-AUTOPSY if you have a natural death that still needs some explaining. Vidal Herrera started the business and he claims that it's a growth industry. In fact, these businesses are opening up in other cities as well. Some of them also procure body parts and offer death-scene cleanup services. (Herrera once recovered semen from the body of a man who'd fallen from a balcony, so the man's fiancée could get pregnant.)

Cyberdeath: Through the Internet, you can watch a funeral live via Webcam, get your ashes scattered in space, contact a mortician, order a casket, make a will, and find out anything you want to know about cemetery culture. You can even watch the front of a funeral home for signs of activity. From "Find-a-Grave" to "The City of the Silent" to "Tomb with a View," it's fairly easy to get information about specific graves or graveyards, instructions in the proper way to do tombstone rubbings, or genealogical studies via cemetery searches. You can also see such subject-specific memorials as the rat cemetery devoted to the memory of pet rats (started by a sixteen-year-old girl who lost two), or go to a temporary Web site that honors a single person's demise.

Unusual Tombs: Two men in Michigan, Dan and Douglas Dudek, started a company that markets pyramids. They claim that the structures can house three hundred thousand human cremains. The base is two hundred feet wide, with the height about equal to a twenty-four-story building. Using polymers, they expect these pyramids to last millions of years. They hope to put the first one somewhere in the Southwest.

Forever Enterprises: Brent and Tyler Cassity not only provide cremations and burial plots in their multiple cemeteries, but also touch-screen biographies of the deceased that can be viewed on a kiosk at the cemetery. They also store these biographies on the Web. At any rate, they feel that who the person was ought to be the focus of a cemetery visit, not some cold memorial stone. A bare-bones package includes twenty minutes of audio and ten photos. The high-end package gets you 115 photos, three favorite songs, an interview with the deceased (before he or she died), and a video of the remembrance party.

Gravestone Artwear Collection: They offer T-shirts with replicas of early American gravestone art. If you can't get to Salem to see the 1688 death's head on Nathanael Mather's grave or the 1794 "Rest in Pieces" carving in Charleston, South Carolina, you can buy the shirts.

YourCoffin: Offers a wooden rendition of the old-fashioned body-shaped coffin with ideas for one hundred and one uses, such as a bookshelf, a wine rack, a Boogie Board, a toboggan, a clothes hamper, and a porch swing. It can also be used as an ironing board or a bathtub. There's no end to what you can do with one of these handy pieces, although I couldn't quite see it as a motorcycle sidecar. My personal favorite was to use it as an isolation unit. I'll have to let Deborah Brown know about that one.

LifeFiles.Com: This company offers Remembrance LifeLegacy Certificates to help families create interactive Web sites to honor their deceased loved ones. When they purchase the certificate from a member funeral home or cemetery, they get the information they need to go on the Internet and create a private site. It's a living memorial with a guest book in which people can write down their memories of the deceased. "This allows the bereaved to build relationships with surviving family members," says LifeFiles.com CEO, Michael Platner. After two years, the site can be renewed on an annual basis.

Coordinated Cremations: Robin Wall, a cybermortician and "direct disposer," offers cremations for $995 on his Web site, icremation.com. Having organized a network of independent cremation providers, he offers to transport a body from anywhere in the country to the nearest place of cremation, place it in a box, and have it cremated. Potential clients need only fill out his online form to get a simple, dignified arrangement.

"I was a doctor's son by birth," he said, "a funeral director by trade, an ambulance operator by accident, and a direct disposer out of necessity." In the business for over twenty years, he found himself in a dilemma. At one time in New Mexico, to get licensed, people could participate in on-the-job training for fifteen years in place of going to school. Yet only a year before he was to achieve this goal, the corporate funeral chains pressured for a new law that left him out in the cold. On-the-job training was no longer a sufficient criterion, but he'd put his whole life into it. A firm believer in the independent family-owned provider, Wall found a way to continue his trade by calling himself a licensed mortician rather than a funeral director, and by opening up a do-it-yourself funeral store: his Internet cremation company.

Entrepreneurialism isn't limited to the U.S. In Malaysia—where those in the death business are generally shunned—David Kong Hon Kong is trying to change things. While making arrangements with great difficulty for his deceased father-in-law, he came up with a plan: He bought his own cemetery, and now he offers a one-stop shop for Taoist, Buddhist, and Christian burials. In short, he offers the same services as most American funeral directors, although in Malaysia this practice is quite unusual. Burials take place in his Nirvana Memorial Park, which is the second largest such facility in the world. Although it's thoughtfully laid out according to the principles of *feng shui*, the Malaysians are still pretty superstitious about planning their own funerals. Nevertheless, the park gets plenty of tourists, who arrive to see such things as the mausoleum laser show and the "singing tombstone"—it's shaped like a piano and when stepped on, plays a song.

TWO

Cemeteries,
Tombs,
and Traditions

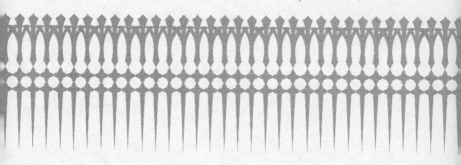

Body Odyssey

Bodies can end up in odd places, and many of us fear losing control over the fate of our remains. Floods and tornadoes disturb cemeteries and any corpse can end up floating downriver or flung across the plains. A man in Louisiana told me about a time in years past when he was out in a rowboat after a flood. He heard a popping sound and turned to see a coffin come up out of the water and float toward him. Then another one popped up to his left, but this one was quite rotted. He looked closer and saw that it was empty. What that told him was that the body had floated out and was in the water somewhere below him. He didn't stick around to wait for it to come up.

Aside from weather events, however, most of us feel reasonably certain that our remains will be handled with respect and will come to rest in the place of our choosing. Yet we can't account for all of the possibilities and sometimes things just happen.

In India recently, a herd of elephants stampeded through the village of Guwahati. One angry elephant pulled a man from a tree and trampled him to death. Then it took him away and carried the corpse around for more than two weeks! No one could get near the beast to remove it.

Then there was the town of Le Lavandou, France, which made dying illegal. Their only cemetery had filled to capacity and the court forbade them to build a new one on the ground of their choosing. So on September 19, 2000, they passed a law that said, "It is forbidden without a cemetery plot to die on the territory of the commune." Until new space is found—which could be several years away—the town's deceased will have to be either shipped out or "hosted" in the vaults of relatives or friends.

A seventy-seven-year-old man and his son-in-law joined a tour group to attend a rugby game in Edinburgh, Scotland. Then they went to a hotel in Glasgow to stay overnight before returning to London. However, during the night, the older man passed away. Panicked, unsure what else to do, the son-in-law carefully dressed the corpse in a shirt, tie, and suit. Then he added a baseball cap, pulling it down over the man's eyes so he could get him onto the tour bus without alerting the others to the man's condition. No one else on the bus knew that they were traveling with a dead man.

I also heard about a woman who put her mother's corpse in the passenger seat of her car, and in echo of the death journey in *Lonesome Dove,* drove one thousand miles from Colorado to Oregon. Although what she did was illegal, she was determined to bury her mother next to her father. Having no idea what the states required, she simply dressed up the corpse, got in, and left. The receiving mortuary went ahead and buried the deceased.

A twenty-two-year-old British man who tried to turn himself into a corpse had a rather shocking encounter. In March 2001, he jumped off the Shakespeare Cliff at Dover, England. He fell some four hundred feet but survived. Rescue workers climbed to the ledge where he lay thirty feet from the bottom of the chalky cliff. The first paramedic to arrive found the victim easily enough, but discovered that he wasn't alone. Next to him lay the badly decomposed corpse of another man who'd made the same leap some time before.

Another strange place that bodies end up is in Dr. Gunther

von Hagens's "Body Worlds." It's a traveling show composed of plastinated corpses posed in various ways to "arouse dialogue." Some people are shocked, others fascinated by the spectacle of a man playing chess with his brain exposed or a woman with an open belly that exposes a fetus. These were once living people, but after they died, von Hagens replaced the fluid in their tissues with synthetic polymers. If the show keeps going, they could travel the globe.

Into the Ground

People weren't always buried. At first, some corpses appear to have been dumped in garbage heaps, and there is evidence that Neanderthals disposed of their dead in caves. Yet the very first complexes of people gathered together appear to have been cities of the dead. Two hundred and fifty miles south of Baghdad in Iraq, archaeologists recently dug up a five-thousand-year-old graveyard that they believe to have been the largest in Sumer. They estimate that it may hold hundreds of thousands of graves, many of which held coffins made of brick and bitumen. The graves were aligned in neat rows, with streets that allowed for visitation, and the dead were obviously buried with their wealth.

In the U.S., the oldest graves are the Indian burial mounds, but European settlers who came in buried their dead around meeting houses. New York City's potter's field started at Madison Square in 1794 and was then moved to other locations. Many stones in New England date back to the mid-1600s.

"Burial" is from the word *birgan*, which means to conceal. All societies seem to have set aside a sacred space for the care of their dead, whether above or below ground, and they generally practiced some kind of rite or ritual. Flower fragments have been found on graves that date back over sixty thousand years, and archaeologists assume that the first burial efforts were to protect the dead from spirits.

Different countries had different practices, but most of those still extant in the United States derive from Europe. The pre-Christian Romans set their dead apart, placing them in aboveground buildings (màusoleums) along the road outside the city, with the wealthiest closest to the living. However, Christians wanted their remains to be near the church, believing they had a better shot at heaven. The first Christian emperor, Constantine the Great, was allowed this privilege and he set a precedent.

The word *cemetery* comes from the Greek word for sleeping chamber. At first, many European cities established graveyards, mostly around religious meeting places, but in light of problems like crowding and disease, the dead were taken out past the city limits. As the available amount of space for burial diminished, alternatives had to be devised, sometimes in a hurry. In Paris, for example, the Cimetiere des Innocents packed in so many thousands of bodies that they literally burst through the foundation walls of neighboring buildings.

In many central European cities, it was impossible to keep expanding the cemeteries, so "rentals" became a solution. To this day in many countries, corpses are generally allowed their spot for a specific period of time, ten to thirty years, and then they are moved to make room for someone else. By that time, they've usually disintegrated. Even for more sophisticated graves, if the family does not pay the fee, the headstone is removed along with the corpse.

In Guanajuato, Mexico, for example, relatives had to pay a renewable fee, and if they didn't, the mummified corpses were removed and placed behind glass in the Museo de las Momias. Most of the bodies were well preserved, down to the skin and nails.

It may also be the case that a single headstone in a cemetery indicates a family plot wherein several members of that family have been interred (known as contumulation). I stumbled across a damaged grave covering in the densely packed cemetery in Old San Juan, Puerto Rico, and when I looked down through the crack of what appeared to be a single grave, I was astonished. Deep down in the ground was a vaultlike chamber, some twelve feet deep and twenty feet wide, that could have held a dozen or more bodies.

Although European burial practices have been traced to the Neanderthals, burning the dead via cremation became a more effective means of reducing a body in the face of overcrowding. It was Christian tradition, with the emphasis on the resurrection of the body, that made burial the disposal of choice in America, but the practice was also influenced by Roman beliefs that access into eternity is prohibited to those who have no body. For the Judeo-Christian community, burial derives from God's statement to Adam that he would return to the earth.

The largest mass graves in the world are generally those from a large number of simultaneous fatalities. They might also be paupers' graves, such as the one in Paris that held over fifteen hundred bodies in a single hole. Graves like that remained opened until filled, which could take several years.

In some religions, the position of the body counts, with the idea that the forces of nature can be harnessed to benefit the dead. Burying a body facedown prevents it from finding its way back up, should some supernatural creature possess it. In Christianity, orienting the feet to the east means assurance of resurrection during Christ's Second Coming (although this originally came from pagan cultures that worshipped the sun). One man in England was buried mounted on his horse, but both were upside down, because he believed that the end of the world would mean that everything would be topsy-turvy.

Cemetery Allure

Originally the land set aside for graves just looked like a flat, open field, but eventually cemeteries became dominated by carved stones and monuments that served as gravesite markers and as personal memorials.

In the U.S., the most impressive cemeteries were the rural garden cemeteries designed in the mid-1800s. Mount Auburn in Boston was the first, in 1831, made possible when James Bigelow persuaded one hundred investors to purchase a piece of property that would have perpetual care. This was a radical new idea, but the feeling of protection it afforded made it spread across the country. Green-Wood in Brooklyn and Laurel Hill in Philadelphia followed a similar plan. These places became fashionable as leisure destinations, and the designer of New York's Central Park, Frederick Law Olmstead, took his inspiration from Green-Wood. Then other cemeteries around the country began to re-create the woodsy look. This meant leaving in the trees, hills, streams, wildflower beds, and any formation that gave people that country feeling.

By the early nineteenth century, a new sense of hope infused views on religion, and an emphasis on the resurrection of the dead diminished fear of the grave. Bodies were no longer viewed as a health hazard, and families visited their deceased to renew their faith in a life everlasting. Death meant passing into something better.

The new cemeteries tried to reinforce that spirit. Cemetery

horticulture was born, followed by the artistic rendering of hope in the form of tombstones, monuments, and mausoleums. Ornamentation was key, as stone carvers imitated the grand sculptures of Europe. For example, the spectacular 1845 monument for seventeen-year-old Charlotte Canda in Green-Wood (designed by her but meant for a deceased aunt) used Gothic arches with rising spires, Grecian urns on pedestals, books of music, and a statue of a girl protected by two ascending angels. Death was the threshold to eternal life, and the Egyptian, Gothic, and Grecian architecture designed to represent this was of museum quality.

Even so, there were still some anxieties about death, notably that one might end up in one of these artistically rendered crypts before one's time.

Taphophobia

Although medical techniques for determining death are better than ever, some people still suffer from an intense dread of going into a coma from which they'll emerge after they're buried. In fact, near the end of 2000, in Kazakhstan, a man did come up out of his grave and actually showed up for his own funeral. He'd been electrocuted and then wrapped in a cloth and placed in a shallow grave. Two days later, he dug his way out, and despite having no clothes, managed to flag down a car. When he got home he found his relatives indulging in his funeral feast.

Now go back three hundred years, when a prematurely buried woman in Scotland had the great fortune of having an unethical undertaker. Planning to return later to steal her jewelry, he buried her in a shallow grave. Then, when he dug her up and tried to remove her rings, she woke up. Not only did she survive the experience (and keep her ring), but she went on to have children and live a relatively long life.

Thus the fear of premature burial isn't totally unfounded. The publication of Edgar Allan Poe's horrifying story of 1844, "The Premature Burial," helped to fuel the belief that such precautions were urgently needed. Blurring the line between reality and fiction, he offered seemingly authentic examples and then posed the dilemma of a narrator with catalepsy who was obsessed with the possibility that he'd suffer the worst of all calamities, to be buried alive.

The story begins with the fact that physicians have documented one hundred such cases and then relates a tale about a woman who was not only laid out for three days after death but was buried only when the appearance of putrefaction had set in. It seemed perfectly reasonable to bury her. Yet when her husband opened the tomb three years later, it was obvious that she had revived in the coffin. Her terrified struggles had knocked it off its shelf and broken it open. She had then used part of the broken wood to strike at the iron door of the vault. In the process, she caught her shroud on the door and finally expired while hanging from her clothing.

Poe's narrator describes just what people most fear about such an experience: "The unendurable oppression of the lungs, the stifling fumes from the damp earth, the clinging to the death garments, the rigid embrace of the narrow house, the blackness of the absolute Night, the silence like a sea that overwhelms, the unseen but palpable presence of the Conqueror Worm . . . " Given his condition, he takes extreme precautions to ensure that his own coffin will be unlocked, air will flow into the tomb, a bell will be fastened to his body, and the door will have a trip device so he can escape. Of course, all of that will only work if his survivors follow his directions. If he awakens in a coffin that has been buried in the ground, without benefit of the bell-rope, then all of his precautions will have been in vain. In short, there may be nothing that one can do to avoid such a fate. To many people during that era, this was truly a horrifying tale. The more such ideas were passed around, the more people looked for ways to avoid being mistaken for dead.

Ancient practices had required a waiting period in which the body began to rot before the relatives would bury it. Three days was common, although it was also accepted to take a few extra precautions. The Romans cut off a finger to ascertain death, and some colonial Americans touched a flame to the tip of a toe. The skin formed a blister that supposedly would burst only if the person was dead. A most decisive method involved cutting the jugular veins, removing the head, or piercing the heart—although if the person were still alive, he wouldn't be after *that*. Even in the 1990s there were recorded cases of mistaken diagnosis of death where the person later returned to life—even to good health. However, if they manage to complete an embalming procedure . . . you're dead.

The worst of these stories involve the later discovery that someone who had revived, post-burial, did not get rescued. During the late 1800s, a man whose crypt was reopened two months after his interment was found lying on his face, the glass on his coffin shattered, and his fists clenched over handfuls of his own hair. A child who had sickened and apparently died was put into a tomb. When it was opened later to inter someone else, her skeleton was found lying near the door.

One premature burial actually resulted in the birth of a baby inside a coffin. In 1901 a pregnant Madame Bobin was diagnosed with yellow fever, and eventually she was pronounced dead. After she was buried, a nurse mentioned to her husband that she wasn't so sure Madame had been dead, so they exhumed the body and found that, after burial, the baby had been born. Since sealed coffins have only enough air to last for about half an hour, the baby suffocated. The father successfully sued the health officers for this doubly fatal mistake.

In the sixteenth century, one man—Francis de Civille of Normandy—seemed destined for live burial. His mother had been buried while pregnant with him, and her husband had exhumed her body to deliver the child by cesarean section. De Civille grew up and went into the army, where he was wounded and then buried for seven hours in a common grave. His servant dug him up to move him and discovered that he was alive. The servant took de Civille to a house to recover and while he lay there, soldiers came in and tossed him out into a dung heap. There he remained buried for three days, but was once again rescued and revived.

Even in 1993 in South Africa, a man who was thought to be dead was placed in a casket and prepared for burial. After two days, he regained consciousness. At first he had no idea where he was, but then realized that he was about to be sealed away. Frantically he knocked on the box and an attendant heard him and helped him to get out, but when he went home to his fiancé, she was so frightened that he was a zombie that she refused to let him speak with her.

Precautions

In earlier times, people who could afford to have someone watch over their corpse sometimes asked that they not be buried for as long as forty days after death, just to make sure they were dead. That meant that an attendant had to watch as that person bloated and slowly rotted. Some taphophobes requested that prior to burial their bodies be pierced in various places, such as straight through the heart, and in 1896 one group formed the Association for the Prevention of Premature Burial. Its members had a deal with certain physicians to perform specific tests that would ensure that they were dead and not just in a deep swoon. The invention of the stethoscope made things easier for physicians, but mistakes were—and are—still made.

Undertakers and casket makers, aware of the possibility of premature burial, invented some interesting devices. One was to write the words "I am dead" with silver nitrate on a piece of glass. This was set or held over a corpse until the body expelled sufficient hydrogen gas to make the words visible.

In some places in Europe, bodies could be placed in small buildings designed specifically for the purpose of letting them rot before burial or allowing them to alert someone if they were not actually dead. Many families actually took photographs of their deceased relatives in these rooms, partly because they were so beautifully arranged. A bell was then attached to the corpse via a cord, so that an attendant could be summoned if necessary. And the bells actually rang from time to time, but this was more often attributed to rigor spasms than to a sudden recovery.

Something similar was eventually included in the design of one type of casket, which Count Karnice-Karnicki patented in 1897. The coffin was sealed, but had a tube running from the ground above to a ball that sat on the chest of the corpse. Any movement inside released a spring and opened up a box at the surface of the grave that then allowed air and light to come into the coffin. A bell inside that box would ring to summon help and a flag signaled the need for a gravedigger. If all of this transpired after dark, the spring action would somehow light a lamp to signal for help. There is no recorded evidence that this invention was ever used, or if used, ever worked, but other patents came out on similar inventions.

Another idea was to equip coffins with speakers and food tubes, so that the parson might make it his duty to go through the cemetery each morning to listen for cries for help. He could then send food down to the unfortunate person while he fetched the exhumation crew. Once the smell of putrefaction was clearly in an advanced state (as it came through the tubes), the tubes could be withdrawn. The inventor, Herr Gutsmuth, tested the coffin himself, and he managed to eat a full meal while ensconced in a coffin, to the satisfaction of spectators.

With the improvement of medical practice around the world and the routine practice of embalming in places like the United States, few people succumb to the terrors of earlier times over the potential for living disposal. Nevertheless, they still want to be buried in some degree of "comfort," and so for those who can afford it, the casket has transformed from a utilitarian container to a bed of luxury. It wasn't until I actually walked through a large casket exhibit that I realized what a variety of boxes there are to choose from—and how vastly expensive they can be.

The Box

While coffins and caskets are essentially the same thing, some people use the word *coffin* to refer to the six-sided body-shaped boxes, while the rectangular boxes are thought of as caskets (from a French word that means a container for something precious). However, they're both used for the transport and burial of dead bodies.

During the nineteenth century, there was a brief fad for coffins made of India rubber, which supposedly retarded putrefaction because they were impermeable. However, over the course of several years the body did indeed rot, and what was left was a smelly, greasy mess which had no place to go.

There have also been porcelain-lined cast iron caskets, pioneered by the Fisk Burial Company, which were quite expensive. Popular from 1850 until around 1890, they had glass viewing ports over the corpse's face. They were welded shut to be made airtight and the degree of preservation was often remarkable.

In the early 1900s the DeCamp Consolidated Glass Company made caskets of leaded glass which were quite heavy and expensive, and they, too, had a preserving effect on the corpse.

People generally make choices from displays at the funeral home. One of the staff at the Frank E. Campbell funeral home told Barbara Walters on *20/20* about an odd occurrence soon after Jacqueline Kennedy Onassis died. Her son, John, and daughter, Caroline, came by to pick out her coffin. After looking around, they both felt a strong sense that one particular model stood out as the only possible one in which to bury their mother. They pointed it out and insisted that they would not change their minds: this was the one. No one had told them, but somehow they had managed to select the very same model of coffin in which their father, President John F. Kennedy, had been buried three decades earlier.

Another place that consumers are now purchasing caskets is on the Internet. Since this is the largest expense of the funeral, consumers are becoming more careful in their choices.

Caskets come in three basic styles, depending on the type of viewing desired.

- Full coach, used when the entire body is viewed

- Half coach, on which the lower part of the lid remains closed while the upper part opens to permit viewing from the waist up

- Hinge cap, which has a removable cover to allow partial viewing of the face and shoulders

They're made of wood, metal, or fiberglass. The cheapest, for just a few hundred dollars, are pressed wood covered with fabric.

There are also kits available for making a casket at home, and some people keep them in their homes as storage containers or coffee tables until needed. I knew a woman who had made a bed inside hers and slept in it (with the lid open), while a man made his into a grandfather clock, saying, "When I stop ticking, they'll stop the clock."

Kentucky woodworker Roy Davis says that he has made a coffin shaped like a steamboat and also plans one to resemble a vintage car from the thirties, with big tailfins and fenders. Davis built his wife's coffin, which she uses as a dollhouse for the same dolls with which she wants to be buried.

Sealers

One thing that seems to bother some people is the idea that things might get into the casket after burial, so to allay these fears, some casket makers offer what they call hermetically sealed or "protection" caskets, which are locked with rubber gaskets to keep out dirt, bugs, and moisture.

However, there have been real problems.

In Bellevue Memory Gardens in Daytona Beach, Florida, mausoleums leaking a red-colored, putrid liquid have brought media attention this past year to the cemetery's apparent mismanagement. Corpses were discovered in the wrong graves, but of more concern was the mix of body decomposition, embalming fluids and rotting caskets that must be bleached regularly. It turns out that the mausoleum construction failed to include a drainage system that would channel fluids to the back instead of the front. This situation had to be rectified.

In another incident, Danell D. Pepson testified before a committee of the U.S. Senate to the effect that a solid copper casket that she'd purchased for the mausoleum burial of her grandmother had utterly failed to perform as promised. The body had eventually decayed, and the embalming and body fluids began to run out of the coffin to the mausoleum floor and then the surrounding soil outside. Pepson believed that the fluids were just rotting vegetation on the walls, so she scraped them away, unaware that they were potentially hazardous to her. When she finally learned the truth, she was understandably distressed. Her own grandmother was fouling the air and soil. Not only that, but her fluids had leaked into the adjoining vault, causing the steel casket of her deceased husband to rust. Both bodies had to be removed and cremated.

It is apparently the case that some hermetically sealed caskets actually act as pressure cookers, causing bodies to bloat and burst open. In the March 2000 issue of *U.S. News and World Report,* Father Henry Wasielewski claimed to have documented how sealed caskets have virtually exploded in the mausoleum vaults, "sometimes blowing the front right off the crypt." Especially in mausoleums where rising air temperatures can have an adverse effect, caskets that have no system of ventilation can spell real trouble.

I recalled that one funeral director had dismissed this as propaganda, but apparently such things happen. After reading this report, I called Lisa Carlson, designer of an end-of-life planning kit and executive director of the Funeral Consumers'

Alliance (FCA), an organization comprised of over one hundred funeral planning societies. They advocate for consumer rights and expose unethical practices in the death-care industry.

Carlson was motivated by her own negative experience to get involved. When her husband killed himself one winter night in 1981, she knew that he wanted to be buried at home. The ground was frozen, so she decided to have him cremated and just bury the ashes. To her surprise, the local funeral homes were charging high fees for this service and she had no money. Then she called the crematory and discovered that they would charge her only eighty-five dollars. So with the help of a friend, she put her deceased husband into a plywood coffin, placed him in the pickup, and drove him to the crematory.

"The reason I knew that I could do that," Carlson said, "is that someone in my writers' groups had written an article about how to bury your own dead in Vermont. If I hadn't read that article, I probably would have bought the funeral home gig, hook, line, and sinker."

Carlson then researched the laws in every state and her book, *Caring for Your Own Dead*, launched the home funeral movement.

About the problems with caskets, Carlson told me that formaldehyde boils at 115 degrees Fahrenheit, and in the South and Southwest, such temperatures are not uncommon. That means that in those areas, problems with leaking coffins are also not uncommon. I'd seen disturbing pictures of dark goo running out from under the cracks in mausoleums, and now I knew the source.

Carlson then described her fact-finding visit to a place in Tennessee affectionately known as "the Body Farm." It was there that she learned more about experiments with the sealer caskets.

As part of the forensic anthropology department at the University of Tennessee at Knoxville, this protected, two-and-a-half-acre wooded plot for the study of decomposing human remains is the only one of its kind in the world. In the 1980s the Knoxville cops dubbed it the Body Farm, as did Patricia Cornwell in a novel by the same name, but resident staff members refer to it as "the facility."

Dr. William Bass III started his field research after he realized how easy it is to make mistakes in estimating when someone has died. He himself had once missed an estimate on a disinterred corpse by 113 years. He thought the man had been dead for a year but the corpse actually dated back to the Civil War. Bass acquired a plot of land and used it to study what happens to the unclaimed cadavers of homeless men, so this place is itself a sort of aboveground cemetery.

To determine time of death, the forensic staff tries to duplicate the various conditions in which bodies are found, and almost three hundred corpses have been used in the studies. They've been locked into car trunks, left in direct sunlight, placed under canvas, buried in mud, hung from scaffolds, locked into coffins, refrigerated in the dark, zipped into body bags, and submerged in water. One may come in headless, another wounded. One may be clothed, another naked. Some are even embalmed. Stages of insect infestation on corpses are studied, along with general exposure to the environment and to animal mutilation.

Carlson had heard that they were doing some studies for a major casket company here, so she had paid a visit. Among the corpses lying about, she soon discovered a row of four black cement vaults set side by side. Tubes channeled out liquid and gas samples, which were measured at regular intervals to determine what happens inside a coffin when the sun heats it up. When Carlson called the casket company to find out more, she was told that the study helped them to improve their products. If that meant continuing to sell sealed caskets, they weren't paying attention. When the caskets containing embalmed bodies heat up to a certain point, it's no longer about keeping things out but about keeping things in.

The bottom line on caskets is this: A casket is meant for the dignified transport of a body prior to burial or cremation. Aside from Evita and a handful of saints, bodies will decompose. So will the casket.

Bodies and Bugs

Speaking of the Body Farm reminded me of that rhyme about the big green bugs. The scientists who've looked into this are called entomologists, and it turns out that the bugs most closely associated with corpses are maggots. In other words, the lowest forms of life go right to work on the highest to send it back to the earth.

The greenish blowflies are the first to arrive, attracted by the noxious odors that a body gives off, and they may settle on the body within ten minutes of a death that took place out in the open. They lay their eggs in convenient orifices such as the eyes, nose, and mouth, so a body found after this infestation is in a most disgusting state. I saw one forensic photo of a man in his bathtub, dead from electrocution by hair dryer. There he was inside his own home, yet he was literally covered with squirmy white worms that nearly obliterated his facial features.

Even the experienced investigator would rather bypass this

stage. The body becomes a host to armies of crawling maggots, which hatch after about twelve hours and then feed on the tissue until they're ready to advance to another stage.

Dr. Bass from the Knoxville facility has made an extensive study of how maggots actually nest in corpses. Because he has cadavers set up all over the field, in many different conditions, he can watch as flies come in and as the eggs hatch and drop off. Often the developing larvae follow a leader, so that a huge group of bugs leaves all at once. Bass has observed that they often depart in long columns and leave a brownish trail, which helps investigators to track them. No matter what their developmental stage, because they have been so thoroughly studied, these maggots provide a host of clues to crime scene investigators.

In one case, a couple of kids riding dirt bikes came across a burned-out car. Inside were the charred remains of a corpse. Investigators came in and removed the corpse, taking it to a lab for an autopsy. Right away they spotted live maggots on the cadaver, but when they removed the skull, they also found maggots inside the brain—now dead and pretty much fried. It appeared from their developmental stage that the maggots had died inside the victim's head some two weeks after he'd died. That proved to be a puzzle, and they surmised that he'd been killed and whoever had killed him had come back and set fire to the car, killing the first stage of maggot infestation. Then more flies came after that and laid their eggs.

Once the flies are done with a decomposing body, the beetles come in—the big green bugs from the song, I guess. They're interested in the skin. The body becomes attractive at various stages to different types of bugs, so there's a fairly predictable natural clock at work, and in some places like Louisiana, red fire ants move in to devour the maggots. Spiders and millipedes do the same. Examining the bug activity can tell a story backward in time about the timing, condition, and context of the death.

In fact, there's even an insect known as the mausoleum fly, although its real name is the Phorid fly. Inside mausoleums, many insects find ways into rotting coffins. The Phorid fly is a gnat that gets through tiny cracks to lay its eggs on the corpses. It's also called the decapitating fly because it lays eggs in the head of a fire ant, causing it to fall off. In mausoleums, the maggots feed off decomposing tissue and the mature flies thus transmit disease, so visitors should be careful to avoid them.

◆ Bodies en Masse

Although most people die individually and have a traditional funeral, it also happens that calamities can produce a lot of bodies all at once in the same place. War, natural disasters, and human error can leave a local community with the overwhelming problem of what to do with the corpses.

THE WAR DEAD

Gettysburg, Pennsylvania, was the site of one of the most significant battles of the Civil War. It raged on for four days in 1863 and resulted in a stunning defeat for the South. As the two armies pulled away to lick their wounds, the entire area became one big horrifying graveyard.

I explored the place with Mark Nesbitt, a former park ranger and historian who now runs Gettysburg Ghost Tours. He had written *35 Days to Gettysburg* about his quest to find the graves of two soldiers—one Union, the other Confederate—who fought across the lines from each other and who both kept diaries of their experiences. One had died, the other survived, and the search had taken him all the way to Savannah, Georgia. The reason for this, he explained, was because of the way the remains were handled after the battle.

"Every soldier that was killed at Gettysburg," Nesbitt told me, "was buried twice. First, they were buried where they fell. Then they were exhumed, and the Northerners were reburied in the National Cemetery here while the Southerners were shipped home." He pointed out that some bodies may still be buried on private property. Not long ago the remains of one dead soldier were discovered near a railroad cut—over a century after he fell.

Before we set out, Nesbitt showed me Gregory Coco's book, *Wasted Valor*, which catalogues numerous descriptions of the decaying, blackened bodies that had lain out on the field for days after the battle. Across thirty-five square miles, it was estimated that some four thousand mutilated Confederate corpses lay waiting for removal. The Union general Meade buried his own soldiers but engaged the town citizens to dispose of the enemy.

The hot weather had assisted the corpses to bloat to nearly twice their original size, which made the eyeballs and tongues distend out of the head. Body parts, shot or exploded off with artillery, added an even more macabre appearance to the field, as did the postures of rigor mortis. One lieutenant wrote, "Several human or unhuman corpses sat upright against a fence, with arms extended in the air and faces hideous with something very like a fixed leer." The sights and smells were so overwhelming that the survivors fell into great fits of vomiting. Within a day, the corpses were bursting out of their clothing, and it was these revolting human horrors that the citizens were forced to bury—people who less than a week before had been calmly tending their farms and enjoying the fruits of a fine summer.

Many of the putrefying bodies were stuck in shallow

trenches, with only a foot of dirt to cover them, and rain eventually disclosed their bones. Others were buried every which way under trees, in yards, near a stream or by a rock. On the trunk of one tree was written "75 REBS BURIED HERE."

I wanted to see some of these places, so we drove along the battlefield and Nesbitt pointed out where men had been shoved quickly into the ground. Eventually the farmers had gotten fed up with running into corpses on their land, he said, and petitioned the governor to do something, so for the Union soldiers land was purchased on Cemetery Hill. To figure out their identities, exhumers relied on letters and diaries found on the rotting bodies.

Nevertheless the Confederates remained in the yards, where property owners were forced to dig shallow graves away from the crops and roll them in. They stayed right there for the next seven years.

In the meantime, the ladies' societies in the South were busy raising funds to have their boys brought home. By that time the bodies were in pretty bad shape, so exhumers tried to collect whatever they could according to the state they were from (based mostly on where they'd fallen) and to ship as many as possible together in boxes. This had been the fate of Thomas Ware, the Confederate soldier killed on the field who had caught Nesbitt's attention. Most of those soldiers were buried together in common graves, marked by a single monument.

Along with new public sentiment that was making cemeteries more inspirational was the desire of America to reclaim and honor its past—especially after the Civil War. Many of the larger cemeteries reserved areas for veterans, and cemetery associations honored the war dead. In 1863 both Antietam and Gettysburg became locations for national military cemeteries and white wooden headboards were placed on more than three hundred thousand graves. War memorials on specific battlefields also engaged the country's attention, and these areas became sacred spots for pilgrimage. Then, when Congress agreed in 1870 to permanently honor the dead, the wooden boards were replaced by matching stones of marble or granite meant to evoke "simple grandeur."

I had that impression as I stood looking at the rows of white markers, but the serenity belied the horror that must have been felt by those who'd collected and disposed of the thousands of mangled, bloated bodies. As a disaster response to mass fatalities, it was clearly hit-and-miss, without much help from national resources.

TITANIC

On April 15, 1912, the unsinkable *Titanic* hit an iceberg in the North Atlantic. The massive ship slowly broke apart and went to the bottom. Some people survived by climbing into boats, but more than fifteen hundred of the passengers died in the icy waters.

That meant the hasty organization of a catastrophe team in the country closest to the disaster, which was Canada. Out of Halifax, Nova Scotia, went the *Mackay-Bennett,* charged with retrieving the corpses. On board they carried ice, embalming fluid, volunteer undertakers, canvas body bags, and over one hundred coffins.

The crew spent almost a week pulling nearly two hundred bodies from the icy graveyard of the North Atlantic. One crewman is quoted as saying, "As far as the eye could see, the ocean was strewn with wreckage and debris, with bodies bobbing up and down on the cold sea." Most of the dead were surrounded by floating ice, and when brought on board, showed damage from the *Titanic*'s descent into the water. Sea creatures had picked at some of them as well. During the night, as the ship went quiet, the moon illuminated the frozen corpses that silently awaited their attention.

Three more ships went out. Altogether, they found 328 corpses. The crew buried more than one hundred people at sea, but every first-class passenger was embalmed and placed in a casket, including multimillionaire John Jacob Astor, who had been badly pummeled by the collapse of the *Titanic*'s funnel. Second- and third-class customers went into canvas bags, and many were paired with weights and recommitted to the ocean.

Although bad weather finally forced the search to end, people on other ships continued to spot bodies floating in the water. One ship bumped a body so hard that it flew out of the water and went several feet into the air. Some corpses still clung to possessions, babies, and even a dog. On May 14, a full month after the disaster, the last official recovery was made—that of the ship's saloon steward, James McGrady.

As the *Mackay-Bennett* returned to Halifax, hundreds of horse-drawn hearses transported the bodies to a local ice rink, now a temporary morgue. Of the 209 corpses brought there, only fifty-nine were claimed. The rest were buried locally, most of them in Fairfax Cemetery. They were interred on a slope, carefully arranged in four rows, and all of the markers bear the same date: April 15, 1912.

D-Mort

Eventually catastrophe management efforts produced a team of professionals from the National Disaster Medical System of the U.S. Department of Health and Human Services that could respond quickly and efficiently to mass fatalities. They're known as D-Mort, or the Disaster Mortuary Response Team, and all of the members are forensic experts of some kind—funeral directors, fingerprint experts, dentists, pathologists, coroners, medical examiners, psychologists, and anthropologists. They arrive quickly to identify bodies and prepare them for the next of kin, and then depart just as quickly.

With disasters like airplane crashes the job is tough because the scene is charred, people may explode into pieces, and the parts of many passengers often commingle. The workers might find a hand with an identifiable ring or a foot in a shoe. To assist their efforts, they use cadaver dogs, which are trained to sniff out body fluids or tissue, including tissue that has decomposed. They can locate corpses under water, rubble, or dirt, and have been instrumental in the quick recovery of missing persons.

D-Mort has a transportable temporary morgue stored at Sky Harbor Airport in Phoenix, Arizona, which offers work stations, autopsy tools, X-ray machines, victim ID tags, medical supplies, and an entire mobile computer network. When the catastrophe overwhelms local resources, they can have it brought in.

Tom Shepardson is the D-Mort national commander. He started it all after what he saw at the Avianca plane crash on Long Island in 1990. There were 159 fatalities, and the immediate need for people who could work with the dead was urgent. He organized professionals to help and then coordinated efforts on behalf of families who sought information or comfort. Once when the only viewable part of a plane crash victim was her arm, Shepardson asked the funeral director to cut a hole in the body bag and put the arm through it so the family, who were otherwise prohibited from seeing her, could go in and hold her hand.

Spotting the need for an efficient way to respond to disasters—specifically to recover the human remains—he continued to organize the forces and in 1992, D-Mort was born. The job of the team was to identify the dead in mass fatalities and to notify families. Oddly enough, their first official task dealt not with a crash or explosion, but with a cemetery.

On January 14, 1993, the Missouri River had flooded and washed out a large cemetery in Hardin County, Missouri. Caskets and bodies were floating into people's yards, and since it was summer, the situation was hazardous. People were also worried that their deceased loved ones would be lost.

When D-Mort arrived, they saw a seventy-foot-deep lake in the middle of the cemetery, and caskets appeared to be floating all over the place. Some were whole, some broken open, some half-rotted. In fact, a total of 769 graves had washed out and the contents had dispersed over sixty-five thousand acres. Caskets, both in and out of vaults, were found as far as thirty miles away, and corpses floated away from where they'd been interred. One by one, those that could be recovered were pulled in. Some people were identified by dental records or from the description of a relative, and most of the modern caskets had serial numbers that had been recorded with the name of the deceased. Even so, the contents of many of the older ones—especially those more than a century old—proved impossible to ID. Bodies and body parts were placed into new caskets and reinterred. It took five months, but finally 607 of the corpses that had been disturbed by the flood were restored to their final rest.

At this point, I had learned about the process of death, the professionals who handle it, and what happens to our deceased flesh on its way to—and inside—the grave. But I realized there was still more to the story, because people don't just die and get forgotten. The monument itself is important, too. As I learned more about what I call "cemetery culture," I began to realize why certain gravestones and monuments are so compelling. Whether it's the story of a life or the story of a death, they provide that human connection.

◆ *Monuments*

Almost as great as the fear of losing control of one's remains is the fear of losing control over one's power of self-definition. French existentialist Jean-Paul Sartre wrote the play *No Exit* to confront us with this very situation. In it, a man and two women who have died find themselves in hell together, and they get to listen in to what their associates remember about them. The man learns that people think he's a coward, which he cannot tolerate, and the women get a similar jolt. There's nothing they can do to reassert their personal definitions, and who they "are" becomes pretty much what others decide.

Not everyone can tolerate this, and in fact some people find *No Exit* to be rather horrifying. Thus, they make their tombstone the place to say something definitive about their personality.

It's the final word, so to speak.

Because grave markers can be both artistic and personal, and because whole books have been written on epitaphs alone, I set out to talk to some monument makers. To my disappointment, I discovered that the art of making a monument is no longer quite the same craft as it once was. These days most stones are machine made, and few people opt for the exquisite carved or molded statuary found in older cemeteries. It's true that fashions change with the times, but cemeteries once were destinations for city folk to commune with nature and enjoy the architecture.

Richard Dickenson grew up in a family in Morristown, New

Jersey, who for over eight decades carved headstones and cemetery monuments. His father had actually made the ledger tablets (large marble sheets covering the grave site) for Thomas Edison and his wife, using Edison's design.

When sandblasting came in during the 1930s, men with Dickenson's skill were phased out. He took up sandblasting, but felt it lacked creativity because it was done with abrasive air pressure and preset rubber stencils. Eventually the stones were finished at the quarries. The only contact people have with monument makers now is through middlemen—a salesperson with a catalog. That's partly why many modern cemeteries have such a uniform appearance.

However, in some places things appear to be changing. What a monument or grave marker looks like is often influenced by the ethnicity or religion to which the surviving family belongs, and some people want an obvious symbol of the deceased's life. It's also about money: The ostentatious wealthy may extend their flashiness into death, and this can result in some magnificent architecture. More and more people these days are taking their burial decisions into their own hands and finding ways to make a personal statement.

I find that there's no cemetery less inspiring than a memorial park. I'd rather see an old-fashioned weeping angel or sleeping lamb than a plaque that's level to the ground with only names and dates, or perhaps a raised picture of a flower or cross. Yet recently I've noticed attempts to revive storytelling traditions even in memorial parks. Many laser etchings of such items as a Harley-Davidson motorcycle or displays of someone's actual oceanfront home provide a visual account of what was meaningful to that person. Others have chosen to have a full version of their story told. One North Carolina resident, for example, designed a twenty-seven-foot-long stone to make room for inscribing twenty-seven paragraphs about himself and his ideas. The Ohio family of a man killed in Vietnam inscribed onto his monument a long poem written by him about his own life.

As I looked into what's available for contemporary monuments, I came across Leif Technologies, a company started by computer expert Deac Manross, which offers a unique type of memorial.

"The trend," he says, "is personalization of the memorial process." For that he's developed a "visual eulogy" whereby a person's life story, with photos, is electronically stored on a computer chip in a stainless steel box with a hinged bronze cover. Weatherproofed and battery-operated, it installs into traditional grave markers and cremation urns. The LCD display screen is approximately five by four inches, and the scrapbook of information and pictures scrolls down in a leisurely way so viewers can fully enjoy it. Twenty pages worth of material is standard, but you can have up to 250 pages. A trucker from Texas plans to have his eighteen-wheel rig buried on his farm, with a Viewology memorial set on top that will describe his many trips.

Other monument handlers foresee some real changes. One company hopes to eventually market hologram technology. That means there may be a visual replica of the deceased that imitates the personality to such a degree that you could go sit in the cemetery and have a "conversation." (I suppose one problem with that is that some people may see this hologram as preferable to having the real person around.)

Along those lines, I've seen entrepreneurs offering cyber-graveyards, which means their customers can place memorials on a computer Web site while they scatter their cremated remains or donate their bodies to science.

That's just as well, since there are interesting things to do with cremated remains. You can pay to have them shot into space, or if you're a hunter, Canuck Sportsman's Memorial Company will load your ashes into shotgun shells and shoot you into the wilderness (or at some wildlife). One widow got the hide of a black bear that her husband's ashes brought down.

Marvel Comics editor Mark Guenwald requested that his ashes be mixed into the ink that printed *Squadron Supreme*. A Yankees fan got her ashes sprinkled around home plate in Yankee Stadium. As long as there are entrepreneurs willing to deal with human remains, there is little limit to ideas on what can be done. However, it's unlikely that many people will opt for what the wife of King Masoleus did in 353 B. C. She dissolved some of his ashes into a potion and drank them down.

 ## Grave Activities

Here I lie and no wonder I'm dead
For the wheel of a waggon went over my head.

Since I was so very soon done for
I wondered what I was begun for.

Here lies Johnny Yeast
Pardon me for not rising

Sayings like these inspired what has become a popular activity known as gravestone rubbing, where people trace tombstone verses and artwork onto paper and collect them from all over the country. To find out just how to do this, I contacted Katie Karrick, also known as "The Cemetery Lady." Calling herself a cemetery historian, she gives tours in cemeteries around Ohio and runs a Web site and newsletter called "Tomb with a View." She has been maintaining her site for over six years and her newsletter has a mailing list of about seventeen hundred.

"I'm on a path," she said about her fascination with graves. "This has come from a lifelong passion of wondering what a symbol means or who did a certain type of sculpture."

Having tromped through many cemeteries around the country and overseas, she has had several singular experiences. One occurred late one September in a cemetery in Dover, England.

"It was getting cool and damp late in the afternoon," she said. "The sun was setting quickly, so by the time I got into the cemetery it was almost dark. Fog was rolling in and it formed a cocoon around me. The unearthly sounds of grazing sheep and cattle were bouncing around in the fog, but I couldn't tell where they came from. I sat down and set my bottle of juice at my feet. No one seemed to be around, so I was enjoying the place, but then I heard someone say, 'This is the direction you have to go in.' I looked over to where the voice seemed to come from, but it seemed to me that to leave, I'd have to turn and go down the hill. Then all of a sudden, I lost my bearings. I had no idea where I was and I started to get afraid. I saw a light in the distance, so I walked toward it, and that's how I found my way out. Then I realized that I'd left my backpack behind, so I had to go back into the cemetery. This time it was really dark, and I looked everywhere for my pack, but just couldn't find it. I stood there not sure what to do. Resigned to the fact that I needed more light, I turned to leave and just then I knocked over the bottle that I'd set down. Right there in front of me was my pack. It was as if I'd been guided out and then been guided back to it."

Since she taught classes in gravestone rubbings, I asked her how to do it and she sent me her list of instructions. Between that list and what I learned from other avid epitaph collectors, I managed to put together a basic idea.

A rubbing is the use of certain materials to obtain impressions of inscriptions, carvings, and designs. Taking impressions from stones or temples originated in 300 B.C. in China, when laws and messages were carved in stone and then transferred onto parchment by rubbing it with wax or dye.

These days, people use such materials as butcher paper, newsprint, tissue, or pellon (fabric lining), although most rubbings are done on basic white paper. The transfer medium is usually a crayon or colored wax.

Briefly, the technique involves taping paper to the stone's surface, making certain that it extends beyond the outside edges. Feel your way around the outside edges of a carving first, and then gently use the crayon or wax to fill in the depressed areas. Keep the strokes uniform in pressure. When you're certain you have it all, carefully detach the paper, roll it up, and stick it into a cardboard tube. Remove all tape from the stone.

While this is a common occupation among cemetery lovers looking for ancestral engravings or renditions of beautiful art, it is in fact illegal in some states, due to the amount of destruction it has caused to the older tombstones. The New England cemeteries have suffered the most, especially those that date back to the 1600s. Whenever pressure is applied to the surface of soft marble or slate, some inevitably scrapes off. Popular stones therefore are most vulnerable.

Gravestone societies offer protocols on "Do's and Don'ts," such as the following:

- It's important to acquire permission.

- Work only on solid stones that are in good condition.

- If a stone has to be cleaned first, one should use only a soft brush and plain water. No detergents or bleach!

- Secure paper with masking tape and make sure no marks are made beyond the paper's edge onto the stone itself.

- Avoid using markers or any other type of pen that may bleed through to the stone.

- Rub gently, with care.

- If a stone sounds hollow or appears to be flaking, leave it alone.

- Avoid using shaving cream, chalk, or any other concoction.

- Don't use nail files or knives to clean inscriptions.

- Avoid spray adhesives, duct tape, or Scotch tape. Use masking tape only.

- Leave none of your materials at the grave site.

- Don't try sophisticated techniques without experienced supervision.

I got all the necessary supplies and went out one weekend to wander through some cemeteries. I wanted something short and to the point, and finally I found one good enough to keep:

She done what she could.

✦ Puritan Art

Personal epitaphs can reveal striking things about the deceased's character, or about how he or she died. Many are amusing, some are stark, others poignant or poetic. The Historical Society of Tarrytown and Sleepy Hollow in New York has a collection of gravestone sketches and photographs that offer a unique glimpse of colonial America. Looking at them, I had a real sense of the many ethnic cultures in the region, and of the ways people once accepted or rejected the fact of death.

In poetic verse, death may be portrayed as a compassionate visitor, a terrible force, or an angelic encounter. For example, the headstone for a fifteen-year-old victim of tuberculosis reads:

> *The pale consumption gave the fatal blow*
> *With lingering pains death found me sore oppressed*
> *Pitied my sighs and kindly gave me rest.*

Ready for more, I continued into New England.

Many of the stones to be found in these places have interesting carvings that identify a specific stone cutter. In other words, they offer a sort of art gallery. In fact, one man whose work dates from 1650 to 1695 was known simply as "the Old Stone-cutter." He was noted for his carvings of rosettes, spades, coffins, and death's-heads.

Like a bed, many graves in the area have both a headstone and a footstone, the latter being inscribed with the initials of the deceased. It was the headstone that provided space for creative flourish, and some of these stones must have taken a long time to carve.

One of the earliest designs in the area was the winged death's-head, depicted in various ways, but suggestive of the transcendence of mortality. In fact, the emphasis on the transience of life is present on nearly every stone from the 1600s that bears a carving. It wasn't unusual for people to die young, so death had to be accepted, and skeletons with scythes and crossbones were also common, as were coffins and hourglasses. Due to the Puritan belief that it was a sin to make representations of the heavenly hosts, angels were not allowed. As that belief eased, stone carvers began to render the soul in flight, which often resembled a winged cherub.

No two stones were alike, but some cutters tended to reiterate their work. For example, William Mumford often rendered rounded devil's heads with carefully carved rows of teeth. The three generations of Leighton family carvers preferred geometric designs, and they also misspelled words. John Homer often used a skull in profile, with crossbones beneath.

The earliest stone markers in New England were made with field boulders until slate from the quarries became the preferred tombstone medium. Eventually marble was used, but at first only the wealthy could afford it.

As fashions changed, common symbols for tombstones included a weeping willow, an urn, a cherub, and broken flowers. Many monuments featured groups of three, such as three steps or three spires, to signify the holy trinity. A broken column or tree without branches symbolized a life cut off before it became all that it could be. All of these symbols, along with the earlier motifs, can be found in the New England graveyards.

I started out in Marblehead, Massachusetts, because several people had urged me to look for a particular monument there. Thus, I set off on a "treasure hunt." The town's oldest cemetery was Old Burial Hill, which I saw from across a pond rising up on a steep mound. The ground was covered with snow, which contrasted sharply with the stiff fingers of dark slate.

Just before I entered the grounds, I encountered a man who told me that the pond was named for Wilmott Redd, a woman who once had lived there. She'd been a victim of the Salem witch trials. He pointed to a spot not far from where we stood where I could see her cenotaph (a monument honoring someone buried elsewhere).

"She's not buried there," he said, "because when they left her body out for relatives to claim, no one did—not even her husband."

I was vaguely familiar with this woman's tragic story. I had often thought about what these accused "witches" had experienced back in 1692 and now I was standing on ground on which this woman had trodden hundreds of times as an ordinary person. How she had been dragged into the Salem fiasco is one of the greatest mysteries of that time.

Redd was apparently an unattractive old woman. Known in town as "Mammy," she had a sharp tongue and the gentry disliked her. Eventually those girls who had accused the women of Salem of witchcraft turned their focus on Mammy, and a warrant was issued for her arrest. She was unaware of what was happening in Salem and Salem Village (now Danvers). Nevertheless, she was taken away.

Mammy was just one victim in a long line of men and women accused by five girls of having dealings with the devil. When three of them were caught dancing in a voodoo ceremony they saved themselves by blaming witches. They put on a show of fits, seizures, and twitches, so the bemused physicians declared them cursed.

Then, by order of the governor of Massachusetts, the Salem Witchcraft Court was formed, one member of which was Magistrate John Hathorne, whose descendent, Nathaniel Hawthorne, authored such Gothic classics as *The Scarlet Letter*

(1850) and *The House of the Seven Gables* (1851). Anyone was fair game for the gallows, and people were denounced, examined, and sentenced, based only on "invisible evidence." At first it was only women, but then men were accused, and even a four-year-old child.

The first official execution was the hanging of Bridget Bishop on June 10, 1692. Then the hysteria spilled into Andover (where two dogs were executed), and reached as far away as Maine. Ultimately 141 people were examined, and twenty were sentenced to die, including Mammy Redd.

No one came to her defense, including her husband, and on September 22, 1692, she was hanged with seven others on Gallows Hill. Afterward, she was examined for "devil's marks," which no one could find, and then tossed over a steep cliff. No one showed up to claim her body, so Mammy Redd was eventually buried in an unmarked pauper's grave.

The cenotaph was a potent reminder that just a few miles away was the place where so many people had been seduced by madness. It was only after the wife of the governor was accused that the trials came to a sudden end. All of a sudden, "intangible evidence" was disallowed.

Leaving Mammy's stone, I went into the cemetery. Old Burial Hill is one of New England's oldest Puritan cemeteries. The first meetinghouse was built here in 1648, and the place was quite different from cemeteries I'd seen across the country. First, it was intensely crowded with stones, all of approximately the same three-foot height, and some of which were falling over onto others. When I stood at the center, just down from a gazebo that crowned the hill, I was struck by the rows upon rows of rounded slate. Most were tall "headboard" type stones, dark gray, a couple of inches thick, and flat on the front and back. It was amazing how they'd stood against the elements for some two to three hundred years, especially on an exposed hill this close to the ocean. The shapes were mostly a rounded half-oval or an oval with "wings." Generally there was a square area on the stone's front for some kind of inscription, and that area was crowned with a rounded arch, called a tympanum or lunette. The wings to the side were for borders of rosettes, figures (death and the devil, perhaps), flower patterns, planets, or just a design such as a banded whorl.

Several unusual epitaphs caught my eye. The marker for Lucey Brimblecom, who passed in 1737 at the age of thirty-nine, indicates that seven of her children—all unnamed—lie there beside her. As I read this stone, a woman nearby told me that children were often seen playing near this grave—but they weren't kids that anyone recognized.

"They're ghosts?" I asked with a smile.

The woman shrugged. "Who knows? I've sometimes felt a tug on my sweater as I walk near here."

I looked away for a moment and when I turned back to say something, she was gone.

Not far away is the grave for Miriam Grose, who died at the age of eighty-one. She left behind "180 children, grandchildren, and great-grandchildren." I guess she lived the good life.

However, the stone I specifically sought was not among these, but at the end of what was known as Minister's Row: that of Susanna Jayne, the wife of a schoolmaster. Her grave was at the foot of the hill, near the pond. Some gravestone enthusiasts claim that this is the most complex and symbolically orna-mented stone in all of New England, and that certainly appears to be the case. It's a tall slate stone that bears a grinning skele-ton wrapped in a winding cloth, crowned by laurel wreaths, and surrounded by a snake eating its own tail (representing eter-nity). The skeleton holds the sun in one hand and the moon in the other. Angels crowd the upper corner while bats take up the lower parts. On top is an hourglass flanked by crossbones. It's fairly gruesome, but in keeping with the Puritan "death is inevitable" theme.

Because of damage to the stone, it is now encased in granite to protect it from falling apart, and behind it is the footstone, which faces in the other direction. On that is the portrait of a winged and smirking woman—perhaps Susanna herself.

Salem is just the next town over, and it exploits the notoriety of the infamous trials with some kitschy shops. Witch T-shirts, bumper stickers, and posters serve as constant reminders of what happened there, and if that isn't sufficient, there are frequent reenactments of the witch persecutions out on the common to lure tourists into the museums.

The oldest recorded burial in the Burying Point Cemetery is 1637, and there are several prominent citizens there. Justice John Hathorne is one of them. When I stood over his grave, I recalled that Nathaniel Hawthorne was so shamed by this ancestor and so affected by the notion that one of the accused might have cursed the family that he'd added a *W* to his surname to distance himself. Justice Hathorne had never expressed regret for his role, although most people rapidly repented. Oddly enough, his family was eventually joined in marriage with the family of one of the accused, and Nathaniel was a descendant of that union.

Although Hawthorne did live in Salem, I knew I wouldn't find his grave. He's buried in Sleepy Hollow Cemetery in Concord, Massachusetts. (A side story associated with Hawthorne is that he got his inspiration for *The Scarlet Letter* from the ornate *A* on the grave of Elizabeth Pain in the King's Chapel Burying Ground in Boston.)

The accused "witches" had no tombstones here. Those who were hanged had been buried on Gallows Hill, now overrun by residences, although the family of Rebecca Nurse did manage one night to secretly retrieve her corpse. Nevertheless, next to the old cemetery is the Witch Trial Memorial, which was dedicated in 1992. It's a cenotaph, but instead of a single marble stone, it's comprised of U-shaped rows of granite benches jutting out from a stone wall. Each bears the name, date, and method of death for the twenty unfortunate victims, and their chilling protests of innocence are inscribed in the stones at your feet. Going to the bench for victim Giles Corey, I sat down. I'd read about him as a kid and have never forgotten the image of what had happened to him.

Refusing to acknowledge the court's right to try him, Corey was sentenced to a terrible punishment meant to force from him a plea of either guilty or not guilty. He knew that if he succumbed, the Crown would take his property, so he refused to talk. The court officials took him out to a field and tied his limbs to stakes in the ground. Then they covered him with a wooden plank and started loading it with boulders. Eventually the weight became so great that it forced his tongue out of his mouth, yet still he would not talk. Whenever he was addressed, he would simply gasp out, "More weight." It took three days, but finally he expired. Apparently before he died he cursed the town.

Taphophiles

I first heard the word *taphophile* while doing cemetery research and discovered that there was a whole network of people around the country who share information about cemeteries and grave sites. Initially I thought a taphophile was a person who was adept at gravestone rubbings, because "taph" seemed to come from epitaph. However, I soon learned that a taphophile is someone who has a strong fascination with the cemetery as cultural artifact—so much so that they might do extensive genealogies, create detailed histories, and memorize large amounts of data about death, funerals, and final dispensations. Some call themselves necrolithologists. Many taphophiles participate in clubs to organize "cemetery crawls" (visits to study and enjoy a specific cemetery), and some even go to cemeteries on their vacations.

Jim Davenport is a good example. He lives in Colorado and he has a passion for a specific type of tombstone carved in the shape of a tree stump. These stones were designed for a fraternal benefit society organized in 1883 called Modern Woodmen of America, and later changed to Woodmen of the World (WOW). Such markers are elusive, so it's quite exciting to find one, and Davenport has made it a goal to photograph and measure as many as he can. "I have found stumps from six inches to ten feet tall, made from marble, limestone, granite, and two beauties made from cast zinc!" His wife soon joined him and they now plan vacations around which cemetery to visit next. The prize discovery thus far is a brown limestone double tree stump made in 1900 for a couple who were killed in a buggy accident in Colorado Springs. "It's ninety-six inches tall, four feet wide, and two feet thick, made from a single piece of stone."

It was the adventures of a taphophile couple, Deanna and Rob from Michigan, that actually inspired me to give this activity a try. Deanna's fascination had grown from trips to cemeteries with her grandmother, but Rob described a more unique experience.

"The incident that hooked me," he said, "happened in Boston when I was twelve, which was in 1965. My aunt took my brother and me on the Freedom Trail walk in downtown Boston. On this walk we visited an old colonial cemetery and I went over to the grave for Paul Revere. I was standing there looking at it when an old woman came up to me and asked me in a kindly tone if I was *sure* that Paul Revere was buried there. I said, 'Well sure, it says so.' Then she said, 'Come with me.' She led me to another spot in the cemetery and pointed out another marker, much smaller than the first, that said PAUL REVERE. She smiled and said, 'Maybe he's buried here instead.' I was confused. I didn't know, but I was fascinated. However, the old lady wasn't done with me. She led me to yet another place in the same cemetery. To my surprise, there was a very short stone, very easy to miss, and all it said was REVERE'S TOMB. The woman then asked, 'So tell me. Which stone is Revere buried under?' That was it for me—I was hooked."

"What we love about cemeteries is discovering the stories that people leave behind for posterity," Deanna said, "what people choose to sum up about their lives on a three-foot-by-five-foot piece of stone." Besides just touring the cemeteries, they also take photographs and collect books about cemeteries, funerals, and death. "The firm criterion for any map purchase," said Rob, "is that it must list cemeteries."

Since they'd taken so many outings, I asked them to describe one that really stood out. The one that Rob recalled gave me a pretty good sense of how persistent taphophiles can be.

They'd looked up some cemeteries in Michigan in which famous people were buried and decided to look for baseball player Norm Cash. Supposedly he'd been buried in Brighton Cemetery, so they put that at the top of the list. They added others, such as Charles Lindbergh's mother, and then set out. Since it was winter, they anticipated some trouble, but the snow on the ground proved to be the least of their concerns.

"In Brighton, we stopped at a gas station and bought a map of the town," said Rob. "I asked the guy about local cemeteries and he directed us to one about a mile away. This was Fairview Cemetery, not Brighton. We went anyway and did some looking around for old Norm, but couldn't find him." They then went to a different cemetery nearby and read every single stone, but still no luck.

"Disappointed, we decided to forget about Norm and try our next stop. We drove out to Pine Lake, where Mrs. Lindbergh was buried. Since it was supposed to be a flat marker on the ground, I knew we'd have to get out and walk. I stopped the car and shut off the engine. Then I opened my door and turned to get out, and right there was a big black marker not thirty feet away that said NORMAN D. CASH."

Another story showed me the allure of serendipity for many taphopiles: the expeditions are like treasure hunts. Deanna and Rob had gone to a cemetery in the middle of Ann Arbor to find Sarah Power, a suicide, as well as the deceased wife of legendary football coach Bo Schembechler. They found Power easily enough, but after driving around for a long time, they couldn't locate the other grave. However, a star-shaped monument on the side of a hill drew their attention, so they decided to climb up and take a look. Once they were that high, they continued to the top, and to their delight, there was the elusive stone for Millie Schembechler.

Since I was in Ann Arbor, I decided to be a taphophile for a day and see what it was like. I could just duplicate what Rob and Deanna had done. First, I did some research on the cemetery itself, known as Forest Hill, and found out that it had been modeled after Brooklyn's Green-Wood Cemetery, so that meant it was one of the rural garden cemeteries with lots of hills and winding roads.

The first grave dated back to 1859, and a matched pair of white Arabian horses pulling a hearse had led the funeral procession. That rather magnificent display had made the cemetery, with its Gothic stone arches, something special to the community. A sign inside indicated that this was one of the largest private cemeteries in the country, with about ninety-eight thousand graves within. Now I understood why it was such a victory for them to have located two graves without having a clue where they were located.

I drove around and kept my eyes open for a tall monument with the name POWER on it. I took the same roads several times, but had no luck. After twenty minutes, I felt somewhat defeated. I couldn't imagine why they thought this was fun. I found it frustrating. Then I noticed an odd-looking grave up on a hill out of which a sapling was growing. I decided to get out and take a look. Walking up, I kept my eyes open for Power and Schembechler. When I got to the sapling, I saw that five pointed stones surrounded it, but I couldn't figure out what they were. Then suddenly I saw the pattern. This was the gravestone that had intrigued Deanna. The five points circling the tree were in the shape of a star. I laughed and went on up the hill to where I knew from their account that Millie Schembechler's grave would be. I had a hit. I was a taphophile!

After that, I started focusing on specific people and their stories, and it was the following graves, cemeteries, and monuments that made the most impact on me.

Necropolis

Colma, California, just outside San Francisco, has the distinction of being the country's only "memorial city," a real necropolis with many more "citizens" dead (over one million) than alive (about fifteen hundred). That's because in 1902 the San Francisco City and County Board of Supervisors passed a law that prohibited further burials in the city. Not only that, but Laurel Hill and Calvary Cemeteries, along with smaller grave-yards, were required to exhume their occupants and move them south. It was probably the most extensive mass transfer of human remains in the history of this country. In addition to the seventeen human burial grounds, there's also Pet's Rest Ceme-tery, which contains over thirteen thousand animal remains, in-cluding monkeys, cheetahs, and goldfish. Many taphophiles told me that Colma is their ultimate vacation destination.

The Big Easy

My own choice of a "city of the dead" is New Orleans. They practice the European tradition there of reusing a tomb, so there's much more handling of human remains. With over forty cemeteries in the general area, and with gruesome tales about bodies popping out of crumbling tombs, it offers every-thing from mystical voodoo rites to ghosts. One guy described how he'd come across a female corpse falling out of a vault in St. Louis Number One and had noticed that the right foot was missing. I'd also heard from several natives about remains that could be seen inside broken graves and coffins that were left in a Dumpster.

The most modern of the cemeteries there is Metairie, which also has the most impressive architecture. Located just off the highway coming from the airport, it really does look like a small city. It contains a pyramid with a sphinx, an Islamic tem-ple, and a ruined castle, along with the "tallest privately owned monument in the United States"—the sixty-foot granite Mori-arty monument.

I like Josie Arlington's tomb, which is the statue of a girl holding a wreath and standing outside a bronze door with her hand poised as if ready to knock. Legends in New Orleans abound about this monument, such as the one my friend Mary told me.

"When Josie was sixteen," Mary said, "she went out with her boyfriend one night and her father warned her not to be home late. When she came after she was supposed to, the door was locked. She knocked on it with a flat hand, but he wouldn't open it, so she and her friend went away to live on the streets. The boyfriend made her into a prostitute so they could get some money."

The truth appears to be that Josie was raised by nuns, but when she was a teenager, she met a man who guided her into prostitution. Whatever the case may be, this all happened at the turn of the century when Storyville, the red-light district, was both legal and popular. Josie became a shrewd madam, achieving power, notoriety, and wealth. "Because she had the money," Mary said, "she designed a monument for her grave to remind everyone of the sixteen-year-old girl knocking on the door where she couldn't get in."

Another tomb that Mary showed me had the name CLARKE on it. Some time in the late 1800s, she said, Mr. Clarke decided to get his funeral plans in order. "He went to get his hair cut and bought a new suit, and then he went to see the funeral director to pick out his coffin. Not long afterward, he went to check to make sure the coffin was ready. They were walking out when he stopped and said he had to check one more thing. He went in and climbed into the coffin and shot himself in the head."

Not far away, off Basin Street, is the oldest cemetery in New Orleans, and the one that seems to draw the most tourists— St. Louis Number One. The tombs date back to the late 1700s, and the most notorious is the reputed burial spot of Marie Laveau, the legendary nineteenth-century Voodoo Queen. It's always marked up with red *X*'s because voodoo practitioners believe that her spirit can be contacted here for good luck. Although the cemetery is locked at night, one person said that voodoo ceremonies still take place inside and that people who live nearby have seen a woman dancing on top of the tomb. Supposedly this is the Voodoo Queen herself.

◆ Ghost Town

Some of the older graves out west offer a different experience. In Arizona there's a town called Jerome that was built up on the side of a mountain. To get there, I drove on up the steep slope of Cleopatra Hill, just below Mingus Mountain. The houses are built on stilts along switchbacking roads that hairpin from right to left.

Founded in 1876, Jerome had grown to a population of fifteen thousand by 1929. Over one billion dollars worth of copper, gold, silver, zinc, and lead had been taken from the mines. It became such a center of opportunity that in 1927 they built a large state-of-the-art hospital that served the entire valley. Then copper prices fell and by 1932, two-thirds of the people had left. Four years later, many of the buildings (including the mortuary) slid down the mountain. The mine closed altogether in 1953, at which time there were fewer than one hundred tenacious inhabitants. They decided to christen Jerome a ghost town.

I stayed at the Jerome Grand Hotel, which had once been the impressive United Verde Hospital. Yet when I asked about a morgue, the staff told me that the hospital had never had one.

"What happened if someone died here?" I asked.

"The undertaker came and picked them up. He looked like Boris Karloff." (So there really *were* undertakers who looked like horror-film stars.) "He'd come in, throw the body over his shoulder, put it into the front seat of his car and drive it down to the morgue in town or into the valley."

Apparently no one much minded. Out west, people just accepted that.

But why take them down to the valley?

"That's where the cemetery was."

That disappointed me. I was expecting some odd little plot along the side of the hill. Then I discovered that in fact Jerome did have a graveyard of its own—Hogback Cemetery—that had been used for only a couple of years. I found it on a hill just outside the town limits and saw that each of the twenty or so graves was surrounded by an iron fence to ward off the free-ranging cattle.

Since there was very little adornment on the small cluster of granite tombstones, I went to the office where the town kept its archives. Historian Alene Alder pulled out a folder about the cemetery, which turned out to be a treasure trove.

"That little cemetery was abandoned," she told me, "because just after they opened it an influenza epidemic hit the town and there were more bodies than it could possibly accommodate. They opened a new cemetery in the valley where it was easier to bury people."

Among the other papers I found in Alene's files were old records of what the undertaker charged. Considering that funerals these days cost thousands, this list was rather amusing.

One bill was printed on the letterhead of furniture dealers, J. P. Scott, Son & McMillan, and the date was October, 1918:

Coffin	$85
Box	$15
Embalming	$25
Hearse	$15
Clothes/Burial robe	$10
Personal services	$5

There were other payments, depending on whether one wanted a carriage, car, or wagon to the cemetery. Twenty dollars for flowers and fifty-one dollars for the lot.

Total charges for one funeral were $203.

Man and Beast

Staying west, I went looking for the grave of Charles Lindbergh on the island of Maui, because I knew that there was something peculiar about it. Although it was a long way to the town of Hana on that island, I knew that a true taphophile would not be deterred.

Interestingly, Lindbergh had planned his own funeral right down to the grave—what's known these days as "pre-need arrangements." He had moved to Maui on the advice of a friend, Sam Pryor, who kept pet gibbons (apes that give off a loud, piercing cry). Lindy bought five acres from Pryor for his eventual retirement, but was soon diagnosed with cancer. He decided that Hana was the perfect spot to be buried and purchased a double-wide plot at the Palapala Hoomau Congregational Church because the small cemetery there overlooked the ocean.

As death approached, he instructed a man to start digging. Each day he wanted to hear about the grave's progress. He had a wooden coffin made and he insisted that he was not to be embalmed. He wished to decompose quickly and return to the stars. Then, in August 1974, he died.

The place is still very much like it was when he selected it for his grave. His headstone, a flat piece of granite two by three-and-a-half feet, is humble and unobtrusive. But one odd thing is this: Next to this man who had won worldwide fame for his risky flight across the Atlantic was a row of graves for apes—compliments of Sam Pryor.

Murder Victims

I've always been interested in victims of violence, so I picked out a few graves that I especially wanted to see.

KATHY BONNEY

In Chesapeake, Virginia, there's a provocative epitaph for Kathy Bonney, who was nineteen when she died: WE TAKE EVERYTHING FOR GRANTED, INCLUDING OUR CHILDREN . . . THEY HAVE NO RIGHTS ANYMORE.

She had written the phrase herself once in a letter, never realizing how it anticipated her own murder.

On November 22, 1987, Kathy was found dead in a canal. She had been shot twenty-seven times. Her own father, Tom Bonney, was arrested and brought to trial. However, the trial took a strange turn.

A psychologist named Paul Dell got involved with Bonney's defense team. He'd read newspaper reports about Bonney's apparent lapses of memory—which included a vague (or faked) amnesia about the murder—and he introduced himself as an expert on multiple-personality syndrome. He suspected that

Bonney might be suffering from this and he wanted to evaluate the man. Dell put Bonney under hypnosis and claimed to have gotten nine other personalities to come through. Not surprisingly, "Satan" emerged, and that's who had killed Kathy.

The jury didn't buy it and found the man guilty, but this was overturned through evidence from another psychiatrist, and the case remains in limbo.

Kathy's friends believed that Bonney had argued with his daughter over a sexually explicit letter from her boyfriend, then shot her and dumped her body in the canal. Those who provided her tombstone had noted the above passage in a letter she'd written to an editor when she'd learned that a man who had murdered his infant daughter had gotten only six months in prison. The words *they have no rights,* echoed her own tragic end, so they used the phrase as her epitaph.

The Smutty Nose Massacre

In New Hampshire, I came across the double grave sites of the victims of a sensational nineteenth-century ax murder not far from Portsmouth. The incident was known as the Smutty Nose Horror, because it happened on Smutty Nose Island in 1873. What's unusual here is that a single stone and epitaph serve both graves:

> *A Sudden death, a striking call*
> *A warning voice, That speaks to all*
> *To all to be prepared to die.*

Louis H. Wagner, a destitute seaman, was aware that three women who lived in a single house were home alone on the island while their husbands worked. Under cover of darkness, he rowed over to the place to rob them. Bungling the job, he woke the women, who began to struggle. He grabbed an ax and hacked both Karen and Anthe Christensen to death. They tried to fight him off, but they lost blood quickly and he finally struck the fatal blows. Then he went looking for the third woman, but she'd jumped from a window and escaped to tell the ghastly tale. The two victims were buried together, unnaturally joined by a most brutal death.

LIZZIE

Speaking of axes, I've always been convinced that despite her acquittal, Lizzie Borden did in her father and stepmother with a hatchet and then disposed of the weapon. She was the chief suspect, but she seemed to have an uncanny ability to foil all attempts to solve this double murder—even after she was dead and buried.

> *Lizzie Borden took an ax*
> *And gave her mother forty whacks*
> *When she saw what she had done*
> *She gave her father forty-one*

That was the rhyme the children sang, but it's not quite right. Altogether there were only about thirty whacks, but those certainly left their mark. It was on August 4, 1892, that the bodies of Andrew Borden, seventy, and Abby Borden, sixty-five, were found in their home in Fall River, in the southernmost part of Massachusetts. Andrew's corpse lay on the living room couch, his face cut by eleven blows—one of which split open an eye. His thirty-two-year-old spinster daughter, Lizzie, found him. The only other person in the house was the maid.

Soon Abby's body was discovered in the guest room. She had been slain with a sharp weapon, possibly a hatchet, inflicting upon her eighteen to twenty blows. Lizzie was arrested and tried, but for lack of evidence (due to the judicial refusal to consider several incriminating incidents), she went free and lived out the rest of her life a wealthy woman, thanks to her father's fortune.

Abby and Andrew were buried in Oak Grove Cemetery, and when her time came, Lizzie was buried there, too, in the same family plot. While it's weird to think that the person who may have murdered two people lies right alongside them, there's another mystery associated with the grave, and that has to do with Andrew's head.

On the day of the double funeral shortly after the murders, the burial was halted. At the grave site, the police informed the mourners that the pathologist wanted to conduct another autopsy, so the bodies were taken away, their heads were removed, and then they were allowed to be buried. Back at the lab, the heads were defleshed so that plaster casts could be made of the skulls. Supposedly the skulls were then placed into boxes and buried in the Borden family plot. Yet there was a rumor that Andrew's skull was not returned. Some thought it was because it held the key to solving the murder in case the murder weapon was ever found.

Because of this rumor, law professor James E. Starrs wanted to open the grave to find out if the skull was actually there. Starrs was well known by then for digging up the bodies of famous people and using new technology to solve old mysteries, such as the case of the kidnapped Lindbergh baby. He'd done it many times without hindrance. But this time was different.

Starrs announced that he wanted to use microscopic analysis to compare chip impressions on a hatchet found a century ago in the Borden home to impressions in the cuts on Andrew's skull. With modern forensic techniques, determining if there was a match ought to be easy. Yet he needed to get access to it.

Using ground-penetrating radar scans over the Borden grave that gives parameters of buried objects, Starrs proved that there had been two different burials. He also surmised that the skulls had been reburied with the bodies, in separate boxes, although he could not be certain without digging up the grave. So he petitioned to do just that, with assurances that he would restore the grave to its original condition. Within days, he believed, he would have an answer to the puzzle of this century-old murder. He felt sure that everyone would want to know.

Yet, to his surprise, locals blocked his efforts. They did not want the historical grave violated. Starrs left town briefly, but then returned to take it to court. Again he was blocked: Since it could not be proven that the heads were actually in the grave, it was ruled, no useful information would come from opening it. This perplexing catch-22 so closely mirrored the legal loopholes that ultimately freed Lizzie that I wonder if she was protecting her secret from the grave.

Unique Cemeteries

Although most stories are attached to monuments or to the way someone ended up in a certain grave, there were also tales about specific types of cemeteries that caught my attention.

SLAVE CEMETERIES

Quite often, black slaves brought in from Africa were deprived of rights to a decent burial, and many were dumped into unmarked potter's fields. In some cases, they were granted a separate section of an established cemetery or were given land for one of their own. For example, a small slave cemetery was found in 2001 on Thomas Jefferson's Monticello property. Because family members were constantly separated, the cemetery became a center and a way to strengthen their fragile community. The Lincoln Cemetery in Gettysburg is a good example, and it's one of the few places where one can find upright antebellum-era gravestones for African-Americans.

Slaves working the cotton and sugarcane fields on Sapelo Island, Georgia, were buried in Behavior Cemetery, established in the early 1800s. With no money to purchase headstones, they generally carved wooden markers, used a cheap composite material known as "tabby," or just grabbed a dinner plate. However, most of the graves, which number in the hundreds, are unmarked, and such is the fate of many of the slave cemeteries in the South.

In the mid to late nineteenth century, blacks were allowed to be buried among whites only in the Northeast. That was because several of the men who had designed the rural cemeteries wanted the place of burial to symbolize equality.

In Cumberland County, Pennsylvania, is the Upper Allen Freed Slave Cemetery. Apparently it was established for slaves who had escaped the South in the 1850s through the Underground Railroad, and it includes fourteen black soldiers who had served in the War Between the States. The dates on the stones range from 1855 to 1862. With no one to care for it, the small cemetery fell into disrepair until a group of Vietnam veterans decided to take it over and clean it up. On behalf of the fallen soldiers, they cleared out the weeds and reset some of the toppled headstones.

Another cemetery for blacks, estimated to be about two hundred years old, was unearthed in 1991 in lower Manhattan.

On September 30, an archaeological team discovered the place just off Reade Street. It was shown on old maps to be the "Negroes Burying Ground," and apparently the place was used for people of African-American descent, paupers, and POWs from the American Revolution. According to a former New York law, people of African-American descent were not allowed to be buried in consecrated ground, such as a churchyard or within city limits. Thus, they had settled on "undesirable" land just outside the city.

During the early years of the medical schools, students had plundered the graves seeking bodies to autopsy. While there was some protest, nothing was done to prevent it. Blacks were considered unimportant, and the unspoken attitude was that students needed the bodies and it was preferable that they pick from society's "lesser" citizens.

The first excavation in 1991, surprisingly, turned up the complete, fully preserved skeleton of an adult male. Lots of personal items were also found, such as brass buttons, bullets, coffin nails, ceremonial beads, and a child's earring. When the digging was done, it turned out to be a six-acre site that had been used for some ten thousand burials, and four hundred bodies were removed for further study.

Yet as news of the excavation spread, African-Americans drew together to protest further desecration of this sacred ground. They acknowledged the value of cultural information, but these were also their ancestors. Mayor David Dinkins appointed a liaison officer who halted everything to give the descendant community more control. A memorial site was developed in which the exhumed remains were reinterred, and since 1993, it has been a registered National Historic Landmark.

Pet Cemeteries

No book on cemeteries would be complete without some mention of the graveyards set aside for our beloved pets. I met Micehele Lanci-Altomare, who'd collected photos of pet cemeteries into a book called *Good-bye My Friend*. She told me that that the oldest known cemetery for animals was discovered in Green County, Illinois, by an archaeologist who estimated that they'd been there since around 6500 B.C. She has come across such tales as the man who brings a weekly beer to his dog's grave, the dog that actually picked out its own cemetery plot, and the mongrel that became the official greeter at a pet cemetery in Las Vegas.

"Many people who lose their pets," she said, "feel as if they've lost a child. You can tell from reading the epitaphs how profoundly these animals have affected their lives."

Even the most ancient civilizations show evidence of ritualized animal burials, and Egyptians actually mummified their pets. However, nineteenth-century European graveyards were overcrowded, so it became illegal to bury pets. America prohibited it, too, so those people who owned no property were forced to toss dead animals into the trash or the river. Thus, pet lovers were delighted at the turn of the century when the first official pet cemeteries were established, one in Asnieres, near Paris and one in Hartsdale, New York.

The Hartsdale pet cemetery began in 1896 when a veterinarian offered space in his apple orchard for a grieving friend to bury a pet. Now it contains over seventy thousand animals of all types, including a lion cub. The grounds are beautifully manicured and among the many unique markers is one carved before World War I that cost twenty-five thousand dollars and weighs fifty tons. Buried there are guide dogs, police dogs, dogs that were war heroes, and dogs that have been in television shows. The cemetery offers last rites and full funerals, including silk-lined caskets.

Almost as old is a pet cemetery in Hamilton, New Jersey, that was first established in the 1890s by the ASPCA. Eventually it was abandoned, then bought and reopened by a partnership of three women who host Adopt-A-Pet events and can set up grieving owners with a new pet. What becomes apparent while touring the place is that almost anyone could take out a small-business loan and do the same thing. These are not licensed funeral directors, but people who see a need and a market.

One customer who comes in brings her living pets with her. She plans to cremate each of them as they die so that when her own time comes, she'll have their ashes buried in the coffin with her. She lines up the urns for previous pets on her night-stand and makes sure that people close to her know her wishes regarding their disposal.

There are now over six hundred pet cemeteries in the United States, and I've even come across one that's reputed to be haunted—the Los Angeles Pet Cemetery in Calabasas, California. Many movie stars interred pets here; Rudolph Valentino's Great Dane, Kabar, has been there since 1929.

"If you come close to the grave," someone told me, "you may feel him lick you on the hand."

Many pet cemeteries started as part of another business, such as a veterinary office or a boarding kennel. To see one of these, I went to Country Kennel Pet Care Cemetery in Michigan.

The cemetery is the size of a large yard. From simple crosses inscribed with the name "Vanilla" or "Dawg" to animal-shaped monuments similar to what you'd find in a regular cemetery, the place was a testament to the love that many owners had felt for their trusted companions. Some included brief histories and a few even had the expensive laser-cut stones. I noted one red lantern that contained an eternal flame device.

Mark Jessop, a strong-looking blond-haired man, opened the cemetery in 1975.

"I started it," he said, "because there were no cemeteries for animals in the area, and I love animals." He currently has four Rhodesian ridgebacks, a malamute mix, and a horse. When he first opened the cemetery, he dug the graves himself, each of them about three and a half feet deep.

"In twenty-five years, many things have happened here," he told me. "There was a young fellow I remember who came out every year for ten years to see a dog that had saved him from drowning."

The pets are buried in caskets of varying styles and sizes, although some may be cremated first. "There are plastic caskets that seal up, or the double-walled caskets with a full lace liner and padded pillows and coverlets. We even use steel baby caskets. Some people bury one pet, but elect to cremate a second and third pet and have those remains buried on top of the first pet."

Besides cats and dogs, there are a few unusual pets. "We've got some goats, birds, even a potbellied pig. I think there are a couple of skunks, too. They make great pets, from what I understand."

He recalled one event that was rather unusual. A man and woman came in with their daughter, who had just lost her pet schnauzer. Jessop could see that she was severely mentally impaired, and when it came time to lower the small casket into the ground, the girl grew upset. She didn't understand what they were doing. Just before they sank it into the hole, she reached down and grabbed the casket and ran with it back to the car. Her parents were embarrassed and they went over to try to wrestle the box away from her. She refused to give it up.

Jessop went over to the car and said, "Just take the casket and go home. It's sealed so she can't open it. She needs that right now. When she's ready, she'll give it up."

The people drove off with the girl still clutching the casket. She kept her pet with her the entire night and into the next day. Her parents worried about how long it might take her to get over this. They couldn't keep the dead dog in the house much longer. Then the girl seemed to realize that her dog was not coming out of the box, so she handed it over and finally allowed them to bury it.

As a side business, some funeral directors will conduct services for pets—where permitted—in the owner's backyard. The package may include a viewing, a slate marker, and an obituary, with the casket an optional extra.

Besides cemeteries and backyards, pet owners can set up memorials online, and members of AOL light candles every Monday night for their deceased pals.

International Oddities

Although there are many stories to be told about cemeteries and graves in this country, I have some favorite spots in other places as well.

THE BONE CHURCH

A most unusual boneyard is a church in the Sednec section of Kutna Hora in the Czech Republic, east of Prague. The place is decorated with the bones of forty thousand skeletons that came from the church's cemetery. There are bone-covered bells that each weigh four tons, and there's a bone chandelier hanging elegantly from the ceiling. The altar, too, is made of bones and skulls, as are the candelabras, statues, crucifixes, and a coat of arms.

This unusual church decor dates back to 1278, when an abbot visited the Holy Land and returned with a jar of earth from Golgotha, where Christ had been crucified. He called it holy soil and spread it over the church cemetery. It wasn't long before people from all over Europe opted to get buried in this sacred ground—particularly people who were dying from the Black Plague. They believed that getting into this soil guaranteed ascension into heaven.

That meant, of course, that the place filled up. But there's only so much space on one piece of land, and within forty years there were thirty thousand bodies crowding the small cemetery. In 1400 a chapel was added to the church and bones were taken from the consecrated ground. A monk, half blind, was charged with the exhumation, so he stacked the bones in an ossuary—or chamber for bones of the dead—built under the church. Then, in 1870, the monks commissioned artist Frantisek Rindt to use the bones to "make a pleasing arrangement." He used between fifty and seventy thousand skeletons to design the wall-to-wall decorations all over the chapel and then left his signature, spelled out with arm and finger bones.

THE GRAVESTONE FOREST

One of the strangest-looking cemeteries, especially from a distance, is in Russia. Just north of a town called Yekaterinburg, in Uralmash, a gang of wealthy but deceased criminals stands tall in a series of etchings on black stones. Each tombstone, which cost from eight thousand to sixty-four thousand dollars (depending on how many semi-precious stones were used on it), is about three meters high. On the face of each stone is engraved the detailed image of a gangster, copied from enlarged photographs, who lies buried in the ground at that spot. Generally the grave's occupant will be depicted with images of power, such as classy tennis shoes or the keys to an expensive car—though they hardly seem very powerful where they are now. Nevertheless, the astonishing expense that went into these monuments is supposed to be a statement to rival gangs that the "family" is still strong. There's a lot of money inside the graves, too, because custom dictates that mourners throw it in with the clods of dirt to give the departed some means in the next world.

The Notorious Tenants of Père-Lachaise

Paris is the home of several large cemeteries, notably Montparnasse and Montmartre, but the way Père-Lachaise became *the* place to be buried is an amusing tale. The French Revolution produced so many bodies in 1789 that the city cemeteries were literally bursting through their walls into buildings next door, so the country's leaders had to think of something to do—quickly. For a while, those who had diminished down to a pile of bones were sent to the catacombs, a series of tunnels sixty feet beneath the city. Many millions of bones were transferred and the tunnels were lined with stacks of skulls, femurs, and arm bones. It's now one of the world's largest ossuaries, open to self-guided tours. Be prepared, however. Just before you get to the bones themselves, an inscription over the doorway warns, HALT! YOU ARE ENTERING THE REALM OF DEATH.

People have been lost down there and one explorer failed to return altogether. His skeleton was found some twelve years after he had entered. When in Paris once, I met a person who referred to himself as a "cataphile," and through him I learned that there are people who love to enter the depths and remain in the secret underground rooms with the dead as long as possible.

But back to the story, as related to me in Paris.

Napoleon purchased a park named Père-Lachaise outside the city walls to the east, but Parisians failed to appreciate his solution. They thought it was too far to go to bury and visit their dead. To make the place more appealing, he disinterred the remains of some famous people, such as Molière, and moved them there. That didn't work, and it looked like this beautiful spot was going to sit unused. However, after the popular writer Honoré de Balzac began to "bury" his characters in the cemetery when they "died" in his novels, things began to change. Curious readers went out to Père-Lachaise to see the locations of these fictional monuments and they saw how lovely the grounds were. Soon the park became a popular resting ground, with a diverse range of artistic monuments housing the likes of Oscar Wilde, Frédéric Chopin, and Marcel Proust. Jim Morrison, lead singer of the rock group the Doors, is there, too, but his grave—aside from all the graffiti—is relatively unassuming. Perhaps that's because if rumors are correct, he's not there.

Several hundred cats flit among the crowded tombs of Père-Lachaise, many of which are stone replicas of full-bodied figures that give the place an eerie feel. One man lies flat on his back to remind people that he died when shot down one day in the streets. When I went there, I was surprised to find that the "paths" are actually cobbled roads marked with street signs, but I was happy to get a map that indicated in which divisions a particular person is buried. Some Parisians told me that underground tunnels connect the cemetery with the city, but to get into one, I'd have to find the right tomb with a secret entrance.

Intriguing as that was, I was more taken with the aboveground sculptures.

In Division 6 lies James Douglas Morrison—supposedly. Being a child of the sixties, of course I had to go see it. It is the fourth most visited site in Paris, and is now under twenty-four-hour surveillance to try to stop the horrendous defacing of other monuments that he seems to inspire in fans. There's little chance of missing this small rectangle of cement amid all the gloriously sculpted memorials: from many different angles, graffiti messages point the way to the Lizard King.

It seems that Morrison, a charismatic musician who was known for songs like "Light My Fire" and "Break on Through," had visited Père-Lachaise a week before his death on July 3, 1971, and had remarked that he would like to be buried there. Was this a prophetic comment or the verbalization of a plan?

Morrison had told some people that he fantasized about starting his life over and living anonymously. One way to do that was to fake his death. At the time, he'd been in Paris for several months with his wife, Pamela Courson, trying to pull himself out of an alcoholic depression. On the evening before he died, he'd had a quiet supper with a friend, who thought he looked ill. Then he went home and during the night got up to take a bath. This is where later accounts differ. Whether Pam was in the room or not is anyone's guess. The same goes for whether she persuaded him to take drugs with her. She claims that she heard his last words: "Pam, are you still there?" Then she found him dead in the water. He was twenty-seven.

Pam purchased the cheapest veneer coffin she could find and by the time the Doors' manager, Bill Siddon, arrived from Los Angeles, she claimed that the body was nailed inside. No autopsy had been performed and Pam "could not remember" the name of the physician who had pronounced the cause of death as heart failure and signed the death certificate only the day before. Siddons never thought to have the coffin opened to check.

At 8:30 A.M. on July 7, 1971, Morrison was secretly laid to rest. Four people attended, no prayers were said, and the funeral lasted all of ten minutes——not much ceremony for the Electric Shaman. Maybe that's because he didn't want much, or possibly because, as he himself once said in echo of Mark Twain, "Rumors of my death are greatly exaggerated."

The mystery of who or what is in the undersized grave might have been solved had the descendants of people buried around Morrison had their way. Thinking that Morrison had the thirty-year lease typical of European customs, they pressured the French government to remove him on July 7, 2001. However, when the matter was researched, it appeared that Morrison had a perpetual lease. Thus his remains remain, and it's likely that the mystery may only be solved if his new persona, Mr. Mojo Risin', decides to make an appearance.

Across the cemetery, in Division 89, Oscar Wilde's pompous tomb became as controversial as he did when he was alive. Author of the infamous *Picture of Dorian Gray* and *The Importance of Being Earnest*, Wilde was convicted of homosexual acts and spent two years in prison. He survived that terrible experience by only three years. To prevent access to his body, doctors had advised that before being entombed in Père-Lachaise, Wilde should be immersed in quicklime to dissolve the corpse. For nine years, he supposedly melted away in Bagneaux Cemetery. Yet against all expectation, when he was exhumed, the gravediggers discovered that the quicklime had actually preserved him. Nevertheless he was transferred to his tomb in Père-Lachaise, where a monument had been erected. To honor Wilde, sculptor Jacob Epstein had created an Egyptian winged messenger, but he'd added something that had shocked the public. The statue came complete with rather prominent genitalia. Officials deemed the statue offensive and insisted that the artist make a stone fig leaf to cover the exposed part. With reluctance, he did so, but it didn't stay on long. Someone decided to take a souvenir and hacked away both the leaf and the offending member.

THE DEAD POETS

The Protestant Cemetery in Rome (Cimitero Acattolico) was designed as a place for separating the foreigners from the "true believers." Until 1738, only Roman Catholics were buried in Rome. However, the next closest burial ground was well over one hundred miles away and getting corpses there proved impractical. Thus, the heretics were finally allowed their own cemetery. However, burials could only take place at night and graves could not be decorated with crosses. Just how the English poet Percy Bysshe Shelley came to be buried here is the rather strange tale of an obsessed fan.

Shelley had come to Italy at the invitation of his lyrical colleague, John Keats, who then up and died. Keats was buried in the Protestant Cemetery, and Shelley went to visit. He thought it such a beautiful place that one might fall in love with death. He had no idea then, but in just over a year, he'd be interred there as well.

It was in July 1822 that Shelley set out in his boat with two male friends. A sudden storm blew up and capsized the boat, drowning the three men. Ten days later, Shelley's body washed up on the coast of Massa. Captain Edward Trelawny, a man who had known Shelley only six months but who apparently worshipped him, went to retrieve the corpse, accompanied by Lord Byron. They discovered that quarantine regulations prohibited bodies to be moved, and Shelley had been buried in the very spot on the beach where he was found. Trelawny was unhappy about this, so he came up with a plan. He would cremate Shelley and take the ashes back to Rome.

Commissioning a portable crematorium, he dug Shelley out of the sand, poured oil onto the corpse, and set the wood on fire. As the body burned, it burst open, revealing the heart. In those days, the heart was valued as the seat of intelligence and feeling, particularly in poets. Thinking quickly, Trelawny grabbed the organ. He burned his hand in the process, but he managed to get away with a real prize—at least according to what he wrote in his memoirs. Mary Shelley wanted the heart and eventually she got it, but Trelawny got something better.

Arriving in Rome, he discovered that the grounds next to the grave of Shelley's son were overcrowded, so he purchased two plots in another area. But the second plot was not intended for Mary. Trelawny himself took his place next to the poet, a man he'd barely known. He designed the tomb and the inscription he wrote was surprisingly romantic:

> *These are two friends whose lives were undivided*
> *So let their memory be now they have glided under the grave*
> *Let not their bones be parted*
> *For their two hearts in life were single-hearted.*

KING ARTHUR IN AVALON

Glastonbury is a small town at the foot of a very tall hill in southwestern England known as Glastonbury Tor. It's reputed to be the sacred place where King Arthur is buried. Once surrounded by marshes so that it looked like a magical land arising out of the water, it's now accessible by road.

According to legend, Joseph of Arimathea, the wealthy man who provided the tomb for the crucified Christ, had brought the twelve-year-old Jesus and his mother to this very spot to learn the wisdom of the Druids. It was considered one of the sacred spots of Britain, the entrance to the underworld of the fairies. Then, after the crucifixion in Jerusalem, Joseph returned to establish Britain's first Christian church. Into a well at the foot of the Tor he placed the chalice from which Christ had drank at the Last Supper—the Holy Grail. The church became a great abbey and a center for pilgrimage. Some say that several hundred years later, King Arthur placed his fortress on top of the Tor.

In 1184, a fire destroyed the abbey. Digging through the ruins to rebuild, the monks discovered a stone slab. They dug down nine more feet and found a hollow log coffin. In it were two skeletons. When they exhumed them to have a better look, they discovered a male skeleton that was seven feet long. It had a damaged skull, as if killed by a blow to the head. The second skeleton was female and it still had fragments of long blond hair. A lead cross lying on top of the male purportedly said, HERE LIES BURIED THE RENOWNED KING ARTHUR IN THE ISLE OF AVALON.

Since Glastonbury once had been surrounded by water and since upon his death Arthur was transported to the island of "Avalon," there might be some truth behind the legend. He was on a quest for the Holy Grail, and Chalice Well was there in Glastonbury. In fact, during Easter week, locals claim that the well water turns a shade of red.

No one knows why.

Cemetery Ceremonies

Despite my reticence about stepping on graves, I have always enjoyed the ceremonies and festivals attached to certain graves or to cemeteries as a whole. Some are meant to honor the dead and others are just for a party.

THE POE TOASTER OF BALTIMORE

Since 1949—one hundred years after mystery writer Edgar Allan Poe was buried—a man dubbed "the Poe Toaster" has come annually on Poe's birthday to place three red roses and a half-filled bottle of cognac on the writer's grave. He arrives at night to pay his silent respects and leaves within moments. No one knows who he is or what the cognac signifies, but people speculate that the roses are for the three people beneath the monument: Poe, his wife, and his mother-in-law.

Poe died in Baltimore on October 7, 1849, at the age of forty, after he was found wandering the streets in a delirium. Initially he was buried in an unmarked grave in a small cemetery next to the Westminster Presbyterian Church (now Westminster Hall). The grave went unattended, and around 1860, people grew concerned. A headstone was fashioned with the sentiment HERE, AT LAST, HE IS HAPPY but a train ran off its tracks near the monument yard and smashed the stone. By 1865, a collection was started to pay for a monument. It took nearly a decade to achieve, but finally it was crafted. However, Poe's birthday was erroneously inscribed as January 20 rather than January 19. The committee decided to place the monument in the front corner of the cemetery so that it could be seen from the road, and that meant exhuming the body to move it.

Many years later, in 1913, another stone was made to mark the original grave, but it was placed too far outside the Poe family plot. Although it was moved, no one is sure that it actually marks the correct spot, and this caused enough confusion to start a rumor. Some say that the memorial committee exhumed the wrong body to place under the monument and that Poe still rests where he once was laid. The Edgar Allan Poe Society supports the idea that the sexton in charge of the disinterment had also buried Poe, so he would have known where the man was buried.

At any rate, the Poe Toaster always comes. In 1993, he "passed the torch," and it was assumed that his successor was his son. The tradition continued as always.

Then, on a rainy January 19 in 2001, the Poe Toaster did something different. He arrived as usual around 2:30 A.M. Clad in black, with his collar turned up, he left his items and departed. The small party of people who had gathered to observe the event pressed in to have a look. The bottle was decorated with blue ribbons and red streamers, and bore a note on it that said, "The New York Giants. Darkness and decay and the big blue hold dominion over all." In other words, the Toaster was a sports fan. He had exploited the occasion to express his support of the Giants' football team over the Baltimore Ravens, named after Poe's famous poem "The Raven." They were pitted against each other in the upcoming Super Bowl.

For many, this message was a letdown. The quote was taken from Poe's powerful story "The Masque of the Red Death," and the Toaster had transformed it into a silly cheer. He'd continued with words from "The Cask of Amontillado," to the effect that "The Baltimore Ravens. A thousand injuries they will suffer. Edgar Allan Poe evermore." Many feel that by doing this, the man damaged his mystique. He hadn't done much for his team, either, because the Ravens won the game.

HOLLYWOOD FOREVER

Across the country, another such tradition took a media-style twist in the Hollywood Forever Cemetery. A registered national landmark, the century-old Hollywood Memorial Cemetery was recently rescued from bankruptcy by Forever Enterprises and renamed. The sixty-two-acre piece of real estate had fallen into disrepair, though it contains the graves of notable producers and actors.

Father Massey, the bishop from Arizona, had a story about this place similar to that of the "Poe Toaster," but this time it was Rudolph Valentino's grave that received the yearly visitor—a woman in black.

"She would show up with a rose and a bottle of whiskey," he recalled. "She'd come into the cemetery, place the rose on his crypt, the whiskey on the floor, and then leave. One day around four in the afternoon, we sat out there in the mortuary across from the cemetery to watch for her. Every major television station in Los Angeles had the same idea. We all waited with great anticipation, but as time passed it was clear that she wasn't coming."

The film crew needed the footage for the evening news, so they weren't about to take it lying down. This was a town of actors, after all.

"One of the film crew came over to the mortuary and said to the secretary, 'Would you put on a black dress and a veil, take this rose, and put it on the crypt so we can get it on film?'" Thus, the tradition continued.

The Birthday Party

Recently some cemeteries have been plotting ways to get people to come through the gates. Since graveyards are no longer a typical weekend destination, unless you're a taphophile, cemetery officials have had to think up enticing activities. The 478-acre Green-Wood in Brooklyn, which once competed with Niagara Falls for the number of visitors, has set up tours and even offers a concert based on the music of the composers buried within. Mount Auburn in Cambridge provides many "arboretum"-oriented programs for the public. (It should be noted that funeral directors aren't altogether pleased with this development. One told me about bringing a funeral procession into a cemetery only to see a woman dressed as the Easter Bunny go hopping by.)

I went to Atlanta, Georgia, for a birthday party—thrown for Oakland Cemetery, which had just turned 150 years old. It is a prime example of a southern Victorian boneyard. Some notable people buried here include Margaret Mitchell, author of *Gone with the Wind*, six of Georgia's governors, and five Confederate generals.

Among the activities were an antique car show, a Victorian hat contest, and a scavenger hunt, and it was at one of the displays that I learned about the following Victorian funeral customs:

- A pallbearer was originally someone who carried the black cloth, or pall, that covers the casket.

- Undertakers and mourners wore black to keep away malevolent ghosts.

- Funeral wreaths were to keep the spirit of the dead person within bounds.

- Birdcages and houseplants were draped in black.

- Diamonds and pearls were for deep mourning, which lasted two years.

- Jewelry was made from the hair of the dead.

- Deaths were announced on cards or stationery with black borders.

The day was capped with a large birthday cake and the invitation to everyone there to come again.

Obon

Some cemeteries have festivities on an annual basis, and I learned that the Obon Japanese Buddhist festival is celebrated every year in Maui (as well as in any other place where there's a Japanese community). Such festivals mean worshipping ancestral spirits to ensure the health of the community. Obon is usually held for a week in July or August, and music, dances, and special food and clothing are always part of the activities. In the midst of that week, for three days, the souls of the deceased are believed to return home, and prayers are said to help guide the confused souls who died during the previous year. One of the most popular of the many events is the Festival of the Dead.

This involves cleaning the grave sites of ancestors, and vegetables, fruit, and rice wine are left three times a day for the spirits to consume. Incense fills the air, flowers adorn the graves, and bright red lanterns are hung out to guide the dead. The grave receives two visits: the first is to retrieve the spirit and bring it to the family home, and the second is to take the spirit back to the grave so it can be on its way. The living then participate in the *bonodori,* a rhythmic, repetitive dance designed to soothe the dead. In cities like Kyoto, Japan, the festivities end with burning rafts being released into the river to guide the spirits back to the spirit world.

The Day of the Dead

There's a tradition in Mexico that each person dies three times. First, the moment in which the body stops functioning. Second, when the remains are consigned to the grave. Third, when the person's name is spoken for the last time.

Some people in the U.S. celebrate the Feast of All Saints, or Day of the Dead, every year in November. I'd participated in the decoration of the cemeteries in New Orleans, but I had heard that it was truly a special treat in Mexico, so I went to Oaxaca, the mile-high capital city of the southern state of Oaxaca. Three friends—Lori, Dot, and Michelle—accompanied me.

Each year on October 31, the citizens begin their family festivities for *La Dia de los Muertos.* Like Obon, this festival is meant to assist the souls on their journey back to their graves in the various *panteons.* One custom is to build *alteres de muertos.* Each altar is made with a corn-stalk arch that represents the sky and a platform on which to place items that had been meaningful to the deceased. The family might set out fruit, tamales, mole sauces, bread (decorated specially for this day), water (because the souls are thirsty), mescal (Mexican liquor), a photo of the deceased, chocolate, sugar skulls, candles, coffins, and flowers as offerings (*ofrenda*) to entice the dead. The most important flowers are the yellow and orange marigolds to help light the way.

I talked with a man who was making such an altar near the city square and he told me, "Once the altar is complete, no one may touch it until the dead have had their fill."

I was quite fascinated with the *tapete de levantada de cruz*, or "carpet of sand," which is generally made in a home where someone has died. This is a rectangle of dirt shaped into forms like dancing skeletons and suffering saints, and then painted with colored sand in red, blue, white, and dark brown. It stays in place for nine days and is then moved to the cemetery to take the soul out of the house.

On October 31, at three in the afternoon, the cathedral bells ring to signal the day of the return to earth of the baptized souls of infants (*angelitos*). Legend says that they arrive at one minute after midnight and stay for twenty-four hours, at which time they leave to make room for the older souls to visit. By the end of the next day, the families can partake of the food from the graves and altars, although the man at the altar had told me, "The food now has no taste." The souls take that with them, but they leave behind positive energy.

Although our intention was to witness this event, we ended up participating in a way that made us wonder about those souls who return.

There was a small cemetery, San Felipe, near where we were staying, and November 2 was their day to decorate the graves. We went up that afternoon to watch. Scattered around a walled cemetery about the size of a football field were dozens of families busily arranging flowers and food on the various mounds and monuments. Some people raked the pebbly dirt while others washed down the white marble tombs. Children ran about, playing games, and dogs wandered around sniffing at the melting chocolate.

Michelle came across the grave of an American and beckoned for us to come and look. The inscription on a raised marble tomb told us that Robert V. Hoppes Rose was interred there. He had been a NASA scientist, composer, writer, and poet, and had died at the age of sixty-one.

"I wonder what he was doing here," I said.

"I don't know," said Michelle, "but he obviously has no one to decorate his grave."

"Why don't we do it?" Lori suggested. "I bought some mescal and we could get some candles and flowers."

The idea took hold, and as evening approached, we collected together a loaf of *pan de muertos*, water, mescal, and a candle to put on Robert Rose's tomb. We approached the cemetery gates and were surprised to find that the street outside had been transformed into a carnival, complete with neon-lighted rides, carousels, food booths, and numerous games of chance. People pushed their way through the crowded avenue as vendors hawked pancakes, tamales, and quesadillas. Apparently this was all part of the party.

Then we walked inside the walled necropolis. I was truly astonished at the veritable garden it had become. Men passed me with bundles of marigolds and coxcombs slung over their backs that were nearly as large as they were, and the darkness was fully illuminated by hundreds of *luminarios* (candles in bags) placed around the graves. It seemed like every grave was covered in thick carpets of blossoms. Families chatted with one another as they ate their picnics, with some playing CDs of rock music and others quietly strumming guitars.

We made our way to Robert's grave and started to lay things out. On both sides, we drew curious looks from the Mexican families. A very old woman, around four feet tall with no teeth and wearing a black scarf over her gray hair, asked in Spanish if he was our relative. Michelle explained what we were doing and she nodded and smiled.

"*Como se llama?*" she inquired, meaning "What is his name?"

"Roberto," Michelle told her, whereupon the small lady stepped close to "our" grave, closed her eyes, and for several minutes recited a Spanish prayer. We waited in silent respect until she was done. Then three little girls noticed that our candle had burned out, so one of them lit it. Someone had placed several bouquets of marigolds in the marble vases that were built into this monument, and I was touched by this generosity.

Then a young woman who had been watching us beckoned for us to follow her. She wanted us to see her relatives, who were sitting vigil over the graves of a man who had died at the age of twenty-nine, and his son, dead at sixteen.

"My sister," she said in Spanish, gesturing toward a dark-eyed beauty with a weary expression. I touched my heart and extended my hand to her, and she nodded acknowledgment.

"We feel their souls nearby," the woman explained. "They're always here with us. Death does not separate us from them."

She couldn't imagine why we in the States did not similarly welcome the souls of our dead back home. It was difficult to convey that many Americans fear and feel diminished by death, and at any rate, most cemeteries prohibit such elaborate displays.

I sensed that they felt sorry for us. After seeing this spectacular ceremony to connect with the dead, so did I.

We sat vigil with them for awhile and then I glanced over at our grave. To my surprise, four women were standing there looking at it. I walked over and when I overheard one of them say something in English, I asked, "Are you looking at our grave?"

"Oh," said a middle-aged woman wearing a black sweatshirt decorated with skulls, "are you his family?"

"No, but we saw that he didn't have anyone, so we adopted him."

Her mouth opened in astonishment. "We did that last year," she said. "We came here and decorated like you just did, and we came to see if maybe his family had returned."

"No way!" I responded. I couldn't believe that we had re-peated their gesture so precisely.

They introduced themselves as four women from three West Coast states: Washington, Oregon, and California. We laughed, because we were four women from three East Coast States: Pennsylvania, New Jersey, and New York. It seemed too unique for words that two separate groups of women had stumbled upon this man's grave and had felt moved to be his "family" for the Day of the Dead. Perhaps it was he who'd been guiding us, rather than the other way around.

When we finally left for the night, we wished Roberto a good journey back, feeling sure that four women from the South or Midwest would show up the following year to take care of his soul.

While our culture has worked hard to scour death from our midst and to make cemeteries serene, there's still an underside to all of this that confirms my original feeling: There's some-thing creepy going on.

Funeral directors had assured me that there's nothing morbid in their work; cemetery caretakers dismissed the idea of ghosts; embalmers said it was just a job; and many people insisted that there was nothing to fear. They were wrong. As I discovered, strange and disturbing things can still happen after we're dead, and there's a shadow side to cemetery culture that's not to be missed.

THREE

Whispers and Shadows
in the Night

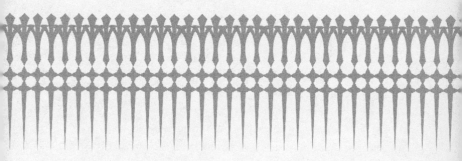

Strange, but True

Apparently the son of Satan, a.k.a. Damian, has died and been buried, because there's a two-foot-tall granite cross for him in Lake Forest Cemetery in Illinois. There were several movies made about this dark spawn, but I'd always thought they were just fiction. In fact, when I first heard about the stone, I figured it was the product of the recent (extremely dangerous) practice among kids of soaking tobacco or marijuana in embalming fluid to get a hallucinogenic high.

Yet there appears to be a genuine grave plot with an inscription to the little devil to the effect that he was born in 1862 and died on November 24, 1932. He's buried in a ravine at the back of the cemetery, his stone leaning forward. Local folklore has it that if he rises again, he will have to walk through water to get anywhere, and somehow that will cleanse him and make him holy. However, what was once a small body of water around the grave has now dried up. From the heat, perhaps?

Laura and Jane live nearby and have visited the stone several times. What they told me is this:

"A local cop told us everyone in the area knows about Damian. He said people leave toy monsters and mess up the grave, but we've seen no evidence of this. What I can tell you is that this grave is very unusual. There are two sections of the graveyard. The first and largest part, south of the ravine, has very elaborate headstones, mostly family plots and mausoleums, and a lot of the graves date back into the 1800s. The newer section, north of the ravine where Damian lies, holds only flat, new graves. I don't think I saw one dated before the latter part of the 1900s. Which means that for many years, Damian was buried alone, far away from anyone else. His stone is crude and amateurish, the only one in the graveyard like it. The jagged rock shows that it was hand-cut, and the writing on the stone could only have been hand-carved. Etched along the side is the word 'ROANIA.' We thought that might be a last name, but it's along the side on an angle, like an afterthought. Maybe it's some kind of a protective mantra? There are also some indecipherable words between the name and the word 'Died,' which incidentally is done in a script that makes the *i* look like a cursive *j*."

They added that they'd heard stories about devil worshippers in Lake Forest, and that some of the mansions had pentagrams and other ritualistic symbols laid into the marble.

So if Satan's son can die, does that mean his soul automatically goes to hell, or can he be redeemed?

This odd little grave brought to mind the medieval Celtic concept of the sin eater, or the person in a community who agrees to eat ritual foods off the corpse of someone who has died. According to the legend, the food represents the dead person's sins, so it's laid on his or her chest and stomach during the wake for the sin eater to come and consume it. Once the food has been eaten, the person is cleansed. However, the sin eater himself has risked his salvation and is subsequently shunned. No one can look on his face or they'll see pure evil. They despise him, but they desperately need him, because only through him can they be saved. (During Ireland's Great Famine, people were only too happy to be the sin eater, because then they'd get fed.)

Another mystical ritual, in Illinois, got a man hauled off to jail. John P. Hawk, forty-three, requested a private viewing of his uncle's corpse outside normal visiting hours. He came and left quietly, but the funeral director soon discovered that his visit involved more than just paying his respects. He had taken his uncle's head. The police arrested him and he underwent psychiatric evaluation. It turns out that he'd founded a business, JPH Health products, which promised to raise people from the dead. Hawk claimed in his flyer that with "bionecrobiology," the soul's energy can pass from the deceased into another person. All that was required was to consume the deceased. He was found not guilty by reason of insanity.

Some people don't want to eat the dead, but they do want to keep a corpse nearby. One man in Arizona lost his twenty-nine-year-old wife to heart failure, and he asked if he could take her body home. Then, for six thousand dollars, he bought an airtight glass container to put her in. This he kept in his living room. He lost a few friends, who were understandably disturbed, but others claim that they just view it as an interesting piece of furniture.

Museums of Death

The business at 6340 Hollywood Boulevard in Hollywood, California, claims to be the only museum dedicated solely to death, and it appears to have the most comprehensive collection anywhere of death-related artifacts and memorabilia. It also has a gift shop. You get there by trekking the "walk of fame" till you reach Bela Lugosi's bronze star. The doors are right there. Within this museum's ten dark rooms you can see the shrouds worn by members of the Heaven's Gate suicide cult (donated by a mortuary worker), along with some of their beds. Other rooms contain running videos of actual deaths, posters of hangings, displays devoted to cannibalism and autopsies, and stained clothing from an electrocution. There's also a serial-killer art gallery, with paintings by such notables as Charles Manson and John Wayne Gacy, and if artwork isn't your interest, there are plenty of photos of crime and accident scenes, along with death shots of the famous.

Across the country is another death-related museum that has some of the most astonishing "oddities" exhibits. The Mutter Museum is located at 19 South 22nd Street in Philadelphia. The College of Physicians of Philadelphia opened the place in 1863 to provide a way to house a collection of instruments and human anomalies that would help to document the more unique aspects of the history of medicine. The curator of this strange resource for medical students and gawkers is Gretchen Worden, who thrives on the bizarre. She couldn't imagine dedicating her life to anything else, and she urges people not to shy away from deformity. The exhibits merely confront us with our own human reality. "It teaches you humility," she says.

On display are the more extreme cases of human abnormalities. For example, there's a row of skulls, numbering over one hundred, from pirates, mummies, murder victims, and people whose skulls were deformed by disease. On the wall are three mummified people who had been eviscerated, and in another area is the skeleton of a man who'd had a disease that caused his muscles to turn into bone. Then there's the "Soap Lady." This obese woman had died and was buried, but was later exhumed. To everyone's surprise, her body had had turned to a brownish soap, or more correctly to "adipocere."

Body Cheese

Now, this is a subject that not many people discuss, but it's part of the decomposition process as well. I first learned about it from a guy who actually goes by the online screen name of "Adipocere." He's dedicated an unusual amount of time and effort to collecting and disseminating information specifically about this substance.

Specifically, the word *adipocere* is French for fatty substance— i.e., that stuff that, when conditions are right, any of us may transform into after death. Generally it depends on having little to no exposure to free oxygen. It took me a few attempts to even read about it, but essentially it's a cheesy substance that forms on the body, and it can be either dry or wet. At first it smells like ammonia, and tends to range from white to beige to brown in color. Later it can take on a sweeter odor. People who die or are dumped postmortem into water are often found with adipocere formation, which tends to preserve their internal organs.

In fact, there was a 1913 murder case in Scotland that was actually solved by adipocere. Two young brothers were discovered floating in a river. When they were pulled up, it was clear that they'd been murdered, but they'd also been in the water for almost two years. It seemed an impossible case, except for the fact that the adipocere that encased their bodies had preserved their internal organs to such a degree that the contents of their stomachs provided a perfect specimen of what they'd eaten for their last meal. This was traced back to a specific region and then to the woman who'd made the meal. It wasn't hard to pinpoint the brutal slayings to the boys' father. Thanks to adipocere, he was convicted.

To distinguish the online historian Adipocere from the substance, I'll refer to him as "A." He seems to know the complete history of humankind's handling of adipocere, from using it for soap and candles to analyzing its chemistry. He claims that possession of true samples is relatively rare, but he knows how to make it.

Born in 1958 in Chicago, A claims that his religious training influenced his unusual hobby. "My interest in all things dead probably began with being a little Catholic kid," A said. "Roman Catholicism is essentially about preparing to die, from the Eucharistic ceremony to the twelve Stations of the Cross to the holy relics."

He first visited a cemetery on his own when he was twelve, after which he developed a fascination with cemeteries, death, and corpses—specifically the stages of decomposition.

"I'm interested in what may happen to the human body after it has died. All things being equal, the greater the time between onset of death and presently, the greater the degree of degradation to the remains."

He can rattle off facts about bacteria, oxygen, and the effects of water in the body the way sports fans list home-run statistics. Then get him going on the effects of various methods of interment.

"Factors determining decay rates," he says, "include ambient temperature of the interment space, humidity, and exposure to air. Generally speaking, the warmer the temperature, the faster the decay rate. Typical burial precludes entry of air, which inhibits aerobic bacteria, but encourages growth of the anaerobic kind. Adipocere formation is thought to be caused only by anaerobic bacteria."

I listened to all of this, but it didn't change my mind about adipocere. Body fat turned to cheese is one of the most hideous things on a corpse that I've ever seen.

The Incorruptibles

Some things never change, including a few corpses. A typical embalmed body may show little sign of necrosis for as long as five months, but stories are told about people who just did not decay—and it had nothing to do with a plastic coating. For example, in 1921 Julia Petta died in her twenties and was buried in Chicago's Mount Carmel Cemetery. A life-size statue of a young woman marks the spot. Legend has it that her mother continued to have dreams about her in which the young woman claimed to still be alive. Finally, after six years of mental torment, the mother obtained permission to open the grave. To everyone's shock, although the coffin had rotted, Julia's corpse still appeared fresh, and no one had an explanation.

Incorruptibility has been thought to be a sign of sainthood. The idea is that some supernatural power keeps these holy people immutable. It may even improve upon their appearance, as was the case with Saint Teresa Margaret, whose fatal gangrene changed from black to a rose color. She even started smelling better, contrary to all expectations.

The story of Saint Bernadette Soubirous is another example. In 1909, thirty years after her death, the Ecclesiastical Court disinterred her remains. According to the reports of those present, she appeared exactly as she had on the day of her death. There were no odors, although the crucifix she held in her hands was covered in mold and her face was a dull white. She was then washed and reinterred. Ten years later, in 1919, her casket was opened, and again she appeared perfectly preserved. A third exhumation was performed some six years later to take relics from the body. She still looked just the same. The surgeon who took the relics was amazed at the preservation, stating that it could not be a natural phenomenon.

And she's not the only one. Such tales can be found wherever sainthood exists, such as when a flood washed away part of a bridge at Avignon, France, in which the coffin of Saint Benezet had lain for five hundred years. The body was recovered and found to show few signs of putrefaction.

Just as astonishing, construction workers digging in the ancient Chinese city of Nanjing unearthed the well-preserved corpse of a sixtysomething bearded man buried five centuries earlier. Experts believed that he'd been a scholar during the Ming dynasty, and they attributed his supple skin and flexible joints to an exceptional casket lined with medicinal herbs.

Even parts detached from bodies may remain fresh, as appeared to be the case with the severed head of England's King Charles I. Although the skin was discolored when it was dug up after 165 years, the musculature was still intact and one eye was as it had been at burial. The hair was thick and black.

However, once these incorruptibles are removed from the environment that has so carefully preserved them (lack of oxygen or bacteria), they do eventually blacken, shrivel, and disintegrate.

Not all saints, it turns out, are incorruptible, and not all incorruptibles are saints. Some just had the right casket. A specific type of metal casket that was made in the 1800s did a remarkable job of preserving bodies. When the Ohio River flooded in 1927, it washed out a casket from a nineteenth-century cemetery. The casket was located and taken to an undertaker for identification and reburial. The remains of the man inside were so well preserved that they had no trouble identifying him as the founder of Henderson, Kentucky. He had been buried 113 years earlier, but his corpse was in such good shape, it might have been only days earlier. Even his clothing and the casket lining were perfectly preserved.

There's a video in the Beacon Museum of Whitehaven, England, of an autopsy performed on a perfectly preserved six-hundred-year-old corpse found in 1981 in nearby St. Bees. Apparently the man died from a chest wound, possibly around 1368. The lead coffin had been soldered shut so it was airtight, and the shroud was soaked in bitumen, then coated in beeswax. Strangely enough, over his chest was laid several locks of a woman's hair.

✦ Shadows and Shades

Telling a story about a corpse is one way to induce anxiety about cemeteries, but it's the corpseless soul that inspires the most fear. Around the world people believe that souls or spirits detach from the rotting flesh and then continue to hover nearby. Some are confused, some angry, some out for revenge. We can't see them, but they have the decided advantage of being aware of us, and many people have told me about a sudden chill or sharp breeze felt near a grave.

Ghost stories go with cemeteries like capes with vampires. Some graveyards are reputed to be haunted because there are regular sightings of a spook, although if modern ghost hunters are right, almost any cemetery can yield an otherworldly encounter. When I was a kid, I used to hold my breath while passing a cemetery in order not to imbibe anyone's spirit.

A GHOST STORY

While I've spent my life as an amateur ghost hunter, I never really thought that keeping a vigil at a cemetery would bring me face-to-face with the filmy essence of a dead person. I mean, no one had actually died in the cemetery, and there was nothing particularly meaningful for them in such places, so why would anyone's spirit haunt it?

Then I met some people who use equipment like digital cameras, motion detectors, temperature scanners, and electromagnetic-field detectors to lock onto the position of an otherwise invisible ghost. To my surprise, they frequently go into cemeteries. In fact, several of them have a theory that cemeteries contain "portals" through which spirits move in and out of our dimension. That means that ghost photography and what they call electronic voice recordings (EVP) are more likely to capture something "anomalous" in a graveyard. In other words, people looking for ghostly manifestations have a better chance here than elsewhere of getting evidence of the spirits.

I had seen these photos—foggy substances, bright round lights, and long strings of illuminated circular objects—so I was eager to try this myself. I went with Rick Fisher of the Pennsylvania Ghost Hunters Society to one of his favorite country cemeteries, near Lancaster. He'd told me about disembodied voices that he had gotten on tape in various graveyards, and I wondered why he did this kind of thing alone at night. One voice had called him by name and another had told him, "Get out!"

The area recently had experienced a serious drought and that night a major storm was on its way: This is a prime condition, I was told, for surges of electromagnetic activity, which in turn gooses ghosts. Rick had a videocam with "night vision" capabilities, while I had a digital camera, a small digital recorder, and a thermal scanner.

We entered a small, enclosed cemetery with around three dozen graves surrounded by a thigh-high stone wall. I took a number of photographs but nothing unusual showed up. Then I had an idea. I went over to the stone wall, aimed my camera away from the graves, and took pictures of the surrounding woods. Bingo! Out there in the trees I had located hovering lights that couldn't be seen with the naked eye. It seemed like they were just hanging out there waiting for us to leave. I recalled the voice on Rick's tape that had told him to "get out."

Rick walked around with his videocam, claiming that he'd recorded several bright orbs that had gone zipping by, so I sat down on the grass and pulled out my tape recorder. I looked around to ensure that there would be no obvious disturbances and then began.

"Does anyone want to communicate?" I asked into the recorder. This was a standard procedure. You invite the spirits to talk. Then I listened closely but heard only the sounds of the chirping crickets. A car went by down the road and then it was silent again. Lightning shot through the distant western sky. Aside from the insects, there was no other noise. After a few minutes, I turned off the recorder so I could play it back. In great anticipation, I pressed the button.

First I heard my own voice asking the initial question, "Does anyone want to communicate?" Then there were crickets, then the car, and then crickets again. Ten more seconds went by and then came a clear voice on the tape that sounded like a young boy's. He simply said, "Yes." This was not Rick's voice or mine, and there was no one else around.

I shivered, sensing that "someone" was quite near, perhaps right behind me, and he wanted to talk. I was about to call Rick over when another voice came on. This one was older, whispery, and I could not make out the gender, but the words were clear enough:

"Why are you doing this to us?"

A chill shot through me. I looked around. Was he talking to me? Why are we doing *what*?

Okay, it was time to leave. Rick and I packed up the equipment and left just as the storm came crashing.

I never returned to that cemetery, and I also never forgot how chilling the experience was.

Reunion

Sometimes ghosts appear to make contact in a less direct manner, as was the case for "Anne." She'd had a very close friendship with a girl in high school, but eventually they parted ways and moved into different worlds. The friend went off to New York and while there she was brutally raped. She then committed suicide. Anne was aware of her death but had never found out where she'd been buried, and that bothered her.

One day Anne was walking through a cemetery near her home and began to think of her friend. She moved into an area that she did not normally go to, but she was looking at the trees and simply found herself in that unfamiliar part of the cemetery. She was about to turn around when she stepped onto a stone that was flush with the ground. Looking down, she was startled. The name on the stone was that of her friend. Finally, she had closure.

Spirited Vaults

On the island of Barbados, in Christ Church cemetery, the Chase burial vault is renowned for its inexplicable history. The first member of the Chase family to be interred was Colonel Thomas Chase's two-year-old child in 1808, soon to be followed by her older sister, the victim of a suicide. Both were placed in heavy lead coffins. Within weeks, the colonel died. When the pallbearers carrying his lead coffin went into the vault, they were startled to see that the other coffins were out of their original places although the seal on the marble slab over the entrance remained unbroken. And it wasn't even as if shifting earth had shaken them out of place. One coffin was lying upside down and across the vault.

They were restored and nothing further occurred until 1816. When the vault was opened, it was clear that all the coffins inside had shifted once again, and quite violently. They were repositioned and the tomb was sealed with cement. However, three years later, the same thing happened, with the colonel's coffin found near the entrance. This time the governor placed his personal seal on the vault and spread sand on the floor to detect intruders. When rumors began of noises within, one member of the Chase family decided to give the vault a thorough inspection. As he and several men chiseled away the mortar, they heard a grating sound inside. Opening the crypt, they discovered one coffin leaning up against the door, while another had been flung so hard across the vault that it chipped a wall. Yet the sand on the floor revealed no footprints.

At a loss to explain it, the owner ordered the coffins to be removed and buried separately elsewhere. The vault still stands empty to this day.

HAUNTED CEMETERIES

Supernatural black dogs with red eyes show up in many world mythologies. Called hellhounds or devil dogs, they appear and then vanish, as if to forewarn of a sudden death. Sometimes a black-robed figure accompanies them, but usually they're alone. Such dogs have been reported patrolling cemetery grounds, particularly slave cemeteries in the South. They frighten most people who see them, but bring comfort to mourners.

I was in one small country cemetery around dusk when a woman warned me not to remain there much longer.

"Why not?" I asked.

"You don't want to be here when the dog comes."

I thought she meant that the cemetery closes and they let loose some sort of guard dog, but that wasn't the story.

"There's a large dog that circles the perimeter," she said. "Some people think it's a wolf. Anyway, the legend is that if you're still in here by the time it circles the cemetery three times, you'll die."

Classic ghost stories often feature the image of a person who keeps going through the motions of how he or she had died. This type of incident has been reported so many times by enough credible witnesses that there's no reason to think it does not happen.

Andrea R. Del Favero, a licensed practitioner of mortuary science, told me the following story about a legendary grave in Totowa, New Jersey. Traveling Route 46 west to the Brower-town Road exit, just past the Passaic River, and getting to Riverview Drive, one eventually reaches the Laurel Grove Cemetery. One night years ago, a girl named Annie was walking home along this road. It was dark and rainy that night, so she could barely see her way. A wind coming up off the river made conditions even worse, but she kept going. She never even heard the truck approaching her from behind. The driver, like Annie, could barely see a thing. He hit her hard and dragged her nearly one hundred feet along the pavement before driving off, oblivious to what he'd left behind. Annie was dead. Her family buried her in the cemetery right next to the scene of the hit-and-run.

The rain that night should have washed away her blood, but

to everyone's surprise, a long red streak remained. It can still be seen on the road, measuring approximately the distance that Annie was dragged.

Many stories about haunted cemeteries involve an actual encounter with the other world. Initially it seems ordinary enough, but then something happens to prove otherwise.

For example I've heard the tale of the man who met a young woman at a dance and offered to take her home. When she felt cold, he offered her his sweater, and then left her off at the house she indicated. The next day he went back to retrieve his sweater, only to be told that the girl he'd dropped off had died years earlier. Unable to accept this, he was shown a picture and only then was he able to accept that he'd been with an apparition. Then he drove away. Eventually he passed the cemetery where this girl had been buried, so he went in to see her grave for himself. There next to the headstone on which her name was inscribed, along with the date of her death, was his neatly folded sweater.

A story from Savannah, Georgia, that began with similarly romantic intentions had a more disturbing ending. This happened not long ago in Colonial Park Cemetery in Savannah's historic district, which is laid out in a series of squares and appears to be saturated in supernatural activity.

It seems that a young blond man used to come quite often to the bar in the eighteenth-century tavern known as the Olde Pink House, which had been built by a man named James Habersham, Jr. The man generally said nothing as he sat there over his beer. A woman who worked in the tavern wondered about his quiet reticence. She watched him every time he came in, and before long she became infatuated with him. Finally she decided to follow him and see where he came from.

So one evening as he got up to leave, she went after him and watched the direction in which he walked. She followed him for several blocks, trying to screw up enough courage to speak to him. Then, just as she was about to say something, he took a quick turn and walked into the old cemetery. Surprised, she stayed back, but watched as he went toward the Habersham family plot. She wondered if he had some connection to this family, if perhaps he was a descendant. He stopped at the iron fence surrounding the aboveground monument and then walked right through it and disappeared.

The girl couldn't believe what she'd seen and thought that maybe it was just a trick of the shadows, so she ran to get a closer look. Although he had seemed as solid as anyone on the street, he simply was no longer there.

Thoroughly shaken by this incident, the woman quit her job and left Savannah.

From a young man named Marcos, I heard about another type of encounter that happened in a small, deteriorating cemetery in Canindé, Brazil.

"The name of the cemetery is Cemitério Municipal de Canindé," he said. "This event took place on a winter day. It was raining torrentially and wasn't going to stop any time soon. Returning from work, a man got off the bus in the wrong place, so he was lost and alone on a dark street. Walking with his head down as fast as he could, he suddenly looked up and saw the old cemetery with its gate ajar. He wanted to leave, but the storm was getting even worse and the street was flooding.

"Then someone called to him, 'Hey, you there!'

"He saw another man about his age, over near the cemetery. Since the stranger did not appear threatening, he stopped.

" 'Come here!' said the other man. 'Stay here until the storm goes away!'

He thought twice about that, but finally he walked over to the other man and went into the cemetery with him.

" 'Terrible night, isn't it?' the stranger said.

" 'Yes, terrible. I don't think I'll get home tonight.'

"They continued to talk together until the rain eased up and the man who'd gotten lost decided to find a way home.

" 'Oh, no problem!' said the stranger. 'I'll stay here. Have a good night.'

" 'Good night, then!' The man left, but then realized he'd forgotten something and came back. 'Excuse me, I forgot to ask your name. What is it?'

"The stranger smiled and said: 'When I was alive, they called me Alberto.'

"That was a strange statement, but a few days later the man who'd been lost came back to look through the cemetery stones to see if any bore the name 'Alberto.' He found one and then described the man to several people in town. They agreed that he'd seen Alberto and then told him that Alberto was indeed dead and buried in that cemetery."

A PRESENCE

Not all encounters are quite so obvious. Gettysburg historian Mark Nesbitt had once lived in a lodge located in the military cemetery there, and he had an eerie experience.

"I lived in a house in the National Cemetery when I was a park service ranger," he recalled. "The chief of maintenance had lived there once and his wife claimed to have heard the sound of babies crying, although their own children were grown. This house had been built on the same site as the original cemetery lodge, where the artifacts were stored that had been removed from the pockets of corpses that had lain all over the field. They hoped that the next of kin would come to claim them. Eventually the building was torn down and the artifacts yet unclaimed inexplicably disappeared.

"So anyway, one night I was downstairs in this lodge watching television when I heard footsteps come down the steps. They stopped on the landing, so I turned to see who it was, but there was no one there. That bothered me. I was so sure someone had come down that I went looking for my housemate. As I went up the steps, it felt like I passed through a cold spot, which seemed strange. I found my housemate in his room, with his door closed. I knocked and asked what he wanted and he didn't know what I was talking about. He said he hadn't come down the stairs."

CADAVER GHOST

Not all cemetery ghosts are benign. On the island of Malta, south of Italy, two cabdrivers had been robbed and murdered near Addolorata Cemetery. The killers had tried removing their heads, but took off before the grisly deed was complete.

One evening sometime afterward, another taxi driver was returning from a run to the airport that took him near the cemetery when he suddenly felt a cold chill. Glancing into the rearview mirror he saw a mutilated man in the backseat. The apparition had two black holes where his eyes should have been.

The driver slammed on his brakes and leapt screaming from the taxi. Then he watched as the apparition slowly rose and floated soundlessly through the locked gates of Addolorata Cemetery.

CONNECTION

Of all the forms an apparition can take, I've only heard of one that involved music. Andrew Saal told me this one. A friend named Sharkey had come up with the idea that they ought to practice their tubas in the cemetery. They agreed to meet at the gates at midnight.

As he got to the place, Andy prepared his tuba. "I tightened the last of the screws," he said, "and then looked up to scan my surroundings. The dim moonlight shadowed the trees and head-stones. A low-lying fog seemed to be flowing in from the woods. I hoisted the tuba over my shoulder and wandered to-ward the woods.

"We found an open area with only a few graves along the perimeter. We respected the dead. No way either one of us wanted to march over someone's grave."

Silhouetted by moon and misty fog, the two tuba players stepped off to the thunder of an imaginary marching band.

"When the imaginary drums rolled off the introduction to the school's fight song, I snapped my tuba to attention and bit into the opening chords. Our high school had chosen for its fight song the 'Notre Dame Victory March.' The pulsing rhythm of the piece that night was almost spiritual. In fact, I swear that I heard the trumpets soaring overhead with that recognizable chorus. Suddenly it hit me. I stopped playing and spun around. And for just a few seconds, lingering on the cool night breeze, I did hear a solitary trumpet playing the final lines. Its sweet per-fect resonance slowly evaporated in the moonlight. For just a second, I felt as if someone somewhere were happy.

"Sharkey stood silently in the fog nearby.

" 'Did you hear . . . ?' "

He nodded almost imperceptibly. Tuba practice was over.

Andy went searching for a likely culprit among other band members, but found no one that had been in that part of town the night before. Then he went back to the cemetery. Walking through, he noticed a military star on one of the gravestones.

"The gray-tinged stone bore the name of James Patrick Sullivan. Born February 8, 1926. Died June 6, 1944. The stonemason had delicately carved the Notre Dame 'ND' symbol onto the headstone. Puzzled, I slowly walked off."

He found a woman, Mrs. Eagen, who lived near the cemetery and who'd been a town resident her entire life. Andy asked her what she knew about James Patrick Sullivan.

She remembered him. The young man had left Notre Dame his junior year to serve in World War II, and he'd died storming Utah Beach on D-Day.

"Almost as an afterthought, Grandma Eagan looked straight into my eyes. 'Did you know that he played trumpet in the marching band?' "

Lingering Spirits

If ghosts do indeed hang out like this in cemeteries, then it seems likely that they might also be in funeral homes. One undertaker told me that "we all have stories," although most would hesitate to admit that.

A funeral director from Tennessee, Dennis Phillips, told me about something that had happened to a colleague. This man was aware of a woman who'd been placed prematurely in a home for the elderly because her son wanted possession of her house. To anyone who'd listen, she claimed that after she died she was going to haunt that son of hers. Years later, when she did pass on, the funeral director picked up her body and set it in the preparation room.

"He went into the office to fill out some papers," Phillips said, "and then went back in. When he opened the door to the prep room, a blast of cold air hit him liked he'd never felt. It went right to the bone. He stood there stunned for a moment. He noticed a curtain flutter slightly, and then it was over. He'd never had such an experience before. But then he heard that within two weeks the deceased woman's son had moved out of her house and wouldn't give a reason why."

More dramatic was the story told to me by an embalmer who was taking a body into what he believed was an empty building. He wheeled the corpse through one room, and since there was sufficient light to see, he decided not to bother with the light switch. Then he noticed movement at the far end of the room, so he stopped and peered down that way. It looked like someone—possibly two or three people—were watching him. Walking back to the light switch, he turned it on and caught a glimpse of someone disappearing into the next room. That was the viewing room, and he knew that no one was supposed to be in there at this time of night. Thinking it might be kids, he went in to investigate. No one was there. He checked the only other door into the room and found it locked. Now he was really mystified. He knew he'd seen someone come in here, yet he couldn't find any evidence. For all he could tell, he was completely alone.

He went ahead and finished his task. Then after checking again and finding no one in the building, he locked up. The next day he mentioned his strange experience to someone else who worked there and she turned pale. "I've seen them, too," she whispered. "They just watch me and then they're gone."

Richard Kramer lives at Hardesty Funeral Home as a caretaker. He's not a funeral director, but he transports bodies to the crematory and runs errands, such as picking up death certificates. It's his belief that although he's the only living person there at night in the funeral home, he's certainly not alone.

"I had someone here who seemed to like to watch television with me," he told me. "When I'd go to bed and turn the TV off, it would be on a certain channel. Then about two o'clock in the morning, I happened to wake up and heard voices downstairs. So I went down and the television was on. I looked and saw that it was on a different channel than I'd had it on and it had dirty movies on. That's not a joke.

"The next night, I turned off the TV, but this time I made a note of the channel it was on when I shut it off. I sat up for about half an hour smoking a cigarette, and then I went to bed. Once again, around two in the morning, I woke up and found the television on. I went down and saw that the channel had been changed to dirty movies on HBO.

"So I said, 'Look, when I watch television, you can watch television, but when I go to bed, you go to bed. The television stays off.' And after that, the television never came back on at night like that.

"At times I did feel that someone was in there with me, but that happens in a funeral home. Strange things occur at night that don't happen during the daytime. For example, I was vacuuming around eleven at night and I turned the vacuum off. I wrapped up the cord and I heard a little voice that said, 'Thank you.' No one was around, so I didn't know who could have been talking to me. So the next night, I decided to see what would happen if I vacuumed again. When I got through and wrapped up the cord, I heard it again, 'Thank you.' It was a dainty, tiny voice. Evidently the noise bothered them."

APPORT

One of the most surprising tales that I heard involved a funeral. A woman was burying her mother and she had left a special ruby necklace on her for the viewing. Several times her husband reminded her to remove it before the casket was closed, but she got caught up in talking with the many people who arrived to show support.

The casket was duly closed and locked, then taken to the cemetery. After a brief service, it was lowered into the ground, and the woman stayed until they had filled in the grave and covered it with flowers. Only at home the next day did she realize that she'd forgotten the ruby necklace. She was heartbroken.

Days later she was cleaning the top of her dresser and to her astonishment, the ruby necklace lay there in a box. It seemed impossible, but it was true. She asked her husband if he had removed the necklace at the viewing, but he denied it. Her only explanation was that her mother somehow had known about her grief and had transported the necklace from beyond the grave. Such items are known among spiritualists as "apports," which is a small material object that a spirit can "bring through" from seemingly out of nowhere. How it got out of a locked coffin and through six feet of dirt is a mystery known only to the dead.

Other Cemetery Entities

In the Highgate section of North London is the thirty-seven-acre Cemetery of Saint James, which dates back to 1839. Around 167,000 people are buried in the fifty-two thousand graves. The most famous occupant is Karl Marx, but Bram Stoker also "buried" one of his characters, Lucy Westenra, in a fictional cemetery based on "Highgate," as it is generally called. Dracula had attacked her and made her a vampire. Out of Highgate she rose at night to hunt for her own prey. It's not surprising, then, that some of the notorious vampire hunters in England claimed that a real vampire has inhabited the cemetery.

Vampires are mentioned in 95 percent of the world's cultures, dating back to the earliest written records. From reanimated corpses to charismatic counts with profound sexual allure, they raise the fear of a creature that has the power to deplete our most precious resource—specifically, our blood. A vampire thrives off the lifeblood of others, taking his or her fill and leaving the victim for dead. It may also be the case that the vampire's attack will transform the victim into another vampire, doomed to face an eternity of thriving by murder and the drinking of blood.

The "vampire of Highgate" was spotted mostly during the evening, and the initial sightings were documented in the 1960s. Some people thought it was a ghost, but according to some accounts, Sean Manchester, the head of the Vampire Research Society, said that a couple of girls told him that they had seen things rising out of the graves. Similar accounts alerted Manchester to the possibility of an undead infestation and he wanted to investigate. One girl, who developed a suspicious case of anemia, actually claimed that an evil figure visited her at night. Only after her room was filled with garlic and holy water, as superstitions dictate, did she improve.

The vampire theory hit British newspapers when it became evident that blood rituals involving animal sacrifices were being performed among the graves. Manchester described how another woman who had seen the entity and who had fallen into a trance had led him to a group of burial vaults. He found several empty coffins, so he and his cohorts lined them with garlic, salt, and holy water, and then placed crosses inside. Apparently the vampire did not return.

However, in 1970 the body of a woman who had been buried in Highgate was found in the cemetery. She'd been beheaded and partially burned, as if someone had taken her to be a vampire. The police stepped in to protect the other interred bodies from such fanaticism. They arrested other vampire hunters who showed up on the grounds, but the reported sightings of something odd hanging around the cemetery continued. There are people today who still claim that a vampire's coffin is secreted away in this venerable old cemetery.

Other vampire tales spread for a time throughout New England, and numerous corpses suffered the indignity of being exhumed and cut open. Henry David Thoreau wrote about such an incident in Vermont—that the family had dug up one of their own and burnt the internal organs.

In 1896, the *New York World* reported that areas of Rhode Island were rampant with belief in vampires that issued forth from the graveyards; six separate incidents were reported near Newport alone. Generally the corpse's heart was taken out (because that's the organ that the vampire supposedly inhabited), examined for evidence of fresh blood, and then burned. When the deaths of a mother and her four children occurred closely together, for example, the villagers exhumed the last victim, removed the heart, and burned it.

This practice was related to the fact that an infected tuberculosis patient typically passed the mysterious illness along to his or her family members. In those days, the illness was called *consumption*—a vampiric term if ever there was one—because people physically deteriorated over a period of time. No one knew the cause, and whenever more than one person in a family died, the community suspected that the already-deceased relative was feeding first on the people it knew best before it targeted others. There is some evidence that Bram Stoker, author of *Dracula*, was influenced by these accounts.

I visited a cemetery where one of these incidents occurred, in Exeter, Rhode Island. It took place in the late 1800s, although it appears not to have been altogether resolved. George Brown lost his wife to TB, and then his eldest daughter died. One of his sons, Edwin, soon grew ill but moved away, and then another daughter, Mercy, died on January 18, 1892. Edwin returned and grew ill again, so George exhumed the bodies of his wife and daughters from the Exeter Cemetery, which was behind the Chestnut Hill Baptist Church. The wife and first daughter had decomposed, but Mercy's body was surprisingly fresh. In fact, the legend says her body was turned sideways in the coffin and blood dripped from her mouth. They assumed that she'd been moving around, which meant she was alive. (Obviously they didn't think of the possibility of premature burial.) They cut out her heart, burned it, and dissolved the ashes in a medicine for Edwin to drink. It didn't help. It wasn't long before he died, and Mercy Brown was saddled with the reputation of being Exeter's vampire. She also shows up as a ghost in this nondescript little cemetery: People around there told me that there'd been reports of blue lights hovering at night close to her grave.

Not far away, a grave in West Greenwich, Rhode Island, bears the epitaph I AM WAITING AND WATCHING FOR YOU. This is for Nancy (Nellie) Vaughn, who died in 1889 at the age of nineteen and was buried in Rhode Island Historical Cemetery #2. Her five siblings supposedly died soon thereafter, which struck up the rumors about her demonic activities. Added to that is the fact that her stone is sinking into the earth and no vegetation can be made to grow on her grave.

One man, Vlad Kinkopf, published an account of his search for this grave, indicating that he came upon it near an abandoned Baptist Church on Plain Meeting House Road just as dusk was falling. He claims that this encounter really happened. He was determined to remain in the small cemetery all night and it wasn't long before he thought he heard muffled noises beneath the earth, located at Nellie's grave. It frightened him, but he stood his ground. Then he saw the faint outline of a woman approaching him, and since he'd heard that the act of offering blood to the dead would get them to speak, he took out his penknife and cut his forearm. She came to him, he claimed, and licked his arm. Then she told him that Rhode Island had more vampires than any state in the union, and that they'd been there since the days of the Narragansett Indians. The proof of this, she said, was that the Indians had tied the hands and feet of corpses, or else had slit their feet to keep them from leaving the burial mounds. Her own immortality derived from a Jewish cult that had discovered the secret of Christ's resurrection, and she had passed that on to others—including a girl named Mercy Brown from Exeter.

Rituals with the Dead

It's not just vampires and ghosts that haunt cemeteries, but practitioners of dark religions as well. I had heard about one cult back in the eighties in New York that stole heads and other parts from fresh graves. While police were sent to investigate the cemetery desecration and to get the headless corpses reburied, the cult evaded capture.

Cult experts speculated that this was the work of a particular sect of the Santeria religion, which is similar to voodoo, but which practices the black arts for nefarious gains. Such people are known as *mayomberos* and there are reputedly no limits to what they will do. They specialize in spells for revenge, necromancy, and murder. To empower themselves they use animal sacrifice, and the apprenticeship requires a peculiar cemetery ritual:

1. First the candidate selects a recent grave that will serve his purposes.

2. Then he sleeps for seven nights under a ceiba tree.

3. After that, he procures a new set of clothing and buries it in the prechosen grave for three weeks while he takes purifying baths.

4. Then he digs up his clothes, dons them, and goes with his teacher back to the tree.

5. Others join them to invoke the spirits of the dead for the initiation.

6. The candidate is crowned with ceiba leaves, which signifies that the dead have possessed him.

7. Finally he receives a scepter made from a human tibia bone that is wrapped in black cloth.

8. Ready now, he returns to the graveyard to perform a ceremony on a recent grave, preferably someone who had been violent or insane in life, because such spirits are more likely to act on orders to go out and destroy.

9. He sprinkles rum in the shape of a cross over the grave and then opens it.

10. He and his companions raise the corpse, remove the head, fingers, toes, tibia, and ribs, and wrap them in black cloth.

11. The new priest then lies on the floor back at home, in the posture of death, and a ceremony is performed to ask the dead parts to do the bidding of the *mayombero*. A negative result means the parts must go back to the grave, and a different grave opened. If positive, another ceremony takes place involving fresh blood and further sacrifice.

To this group of Santeria priests, laws and fines mean nothing. They see themselves as outside the bounds of any code but their own.

I learned that cemetery thefts had occurred in Maui as well. "Hawaiians believed that bones have power," said Debbie Iida, a longtime resident, "and they would steal them from the grave sites. That's why a lot of Hawaiians were buried at night in sand, because sand leaves no trace of the digging. The sand dunes of Maui are full of old bones."

Satanists, too, get into the act, and they generally take the blame whenever there's evidence of a grisly ritual in a cemetery. The grave of General Elisha G. Marshall, who served on the Union side in the Civil War, was violated on the night of the summer solstice in June 2000. He'd been buried in 1883, in a pine casket in Rochester, New York's Mount Hope Cemetery. Sometime between 8 P.M. and 11:30 A.M. the next day, thieves dug down six feet, smashed though the moldering coffin, and took the general's skull. The police found bones scattered on the ground around the open grave, and they surmised that this group believed that grabbing the remains of a corpse during the change of seasons empowered their rituals.

On one of my trips to New Orleans, I decided to visit Holt Cemetery. While people know it as a pauper's graveyard, it's also a spot believed to draw voodoo practitioners in search of body parts and bones.

Believers in voodoo, which is a hybrid of African and Catholic beliefs, claim that the gods speak through spirit possession and that pleasing them results in the reward of a good life. The pantheon of gods, known as *loas*, include the deified spirits of ancestors, and they can inspire either good or evil.

One secret voodoo sect, Bizango, worships the *loa* of the graveyard. To protect their nighttime ceremonies, they have elaborate passwords and rituals. If someone gets in but cannot say the password, he supposedly gets sacrificed or made into a zombie. Those who are quite serious about this art may sleep in tombs to commune with the spirits and thereby gain knowledge and power. At any rate, parts of the dead are thought to heighten these powers.

In fact such a case occurred in Florida in 1997. A garbage collector named Willie Suttle had died and when no one claimed his body, it was taken to a funeral home to be buried as an indigent. No one seemed to notice that Suttle's left hand was missing, but when it washed up onto the banks of the Manitee River, the prints were traced back to Suttle and his body was exhumed. Only then did someone investigate further. Inside him, they found an odd assortment of fabric dolls on which notes were pinned. There were twelve dolls in all, eleven of which were black, and the notes contained what appeared to be chants and curses, along with the names of local funeral home directors. Since the last person to handle the body was Paula Albritton of the Green Funeral Home in Bradenton, she was questioned. Initially she denied knowing anything, but she eventually confessed. She was a voodooist, she claimed, and had used the hand in a religious ritual. This "helping hand" was meant to protect her business and free Suttle's spirit.

However, when detectives told her that they'd be able to find fingerprints on the duct tape wrapped around the wrist from which the hand was removed, she admitted that she'd lied. In fact her mentally ill son, Jimmie Clark, had obeyed voices in his head that told him to stuff the body with the dolls to curse competitors. He was convicted of corpse abuse.

I had this story in mind as I wandered into New Orleans' Holt Cemetery.

Established in 1879, this crowded, seven-acre plot is unlike the many aboveground cemeteries in the area, and in fact the majority of graves are marked with improvised wooden crosses. Though thousands of burials have taken place here, many grave markers have no names, and some of the graves are loaded with weeds and trash. Strangers share caskets as the remains of one occupant of a particular grave are dug up to make room for another. Leftover bones and teeth get bagged and tossed into the new casket.

Off to one side stands a magnificent oak tree hung with Spanish moss, and that's apparently where voodoo rituals take place under the cloak of night. Several people told me that a small group gets together here on a fairly regular basis to draw on the power of the dead.

In the past, the sextons reported that odd things have been buried in the graves; clearing them up might yield a collection of pins and needles—implements common to voodoo. People also take dirt from a relative's grave to sprinkle around the house, and some voodoo practitioners believe that taking a withered flower and a pinch of dirt from the grave of an enemy makes the possessor powerful over that person's people.

On a cold December afternoon, not long before twilight, I walked in through an iron gate to look around. People had dumped off old ironing boards, tires, and broken furniture here, yet I saw attempts at artistic expression as well. I walked past cracked, sunken graves, as well as fresh mounds of dirt that suggested recent burials. Then I stopped near one grave just off the main path. I thought I'd seen something odd and decided to have a closer look. I stepped over to the grave site and bent down.

Sticking out of a heavy-gauge black bag dumped on top of the grave was something white that looked like bone. I nudged the bag with my finger, but it felt solid. The bag was torn a little, so I opened it further to have a look. I still couldn't tell.

I went and found a stick, and again bent over the bag. Digging at the white object, I managed to move it out a little further for inspection. The chill I experienced was not from the overcast skies. There was no doubt about it, this was a bone of some kind. It was fairly sizable, too, about the measure of the top of a femur. Someone had dumped a sack of remains here on top of this grave.

The bag wasn't large enough to hide a corpse, but it could certainly contain someone's half-baked cremains. It might also hold the hasty deposit of exhumed parts.

Looking around, I wondered if anyone was watching. The bag appeared to have been here awhile, and surely others had seen it, but I wanted to take no chances. The sun would be gone within half an hour, and since I didn't know the password to any voodoo ceremonies, I decided that it was time for me to leave.

Corpse Abuse

By far the majority of people who work in funeral homes are professional, caring people, but they're aware that there are some who get into the business for nefarious reasons. One funeral director hinted that there was a secret underground, a network of sick individuals who protect one another. They might even go so far as to "share bodies," i.e., call others when an attractive corpse comes in to indulge in a party.

In fact it was a caterer who described one such gathering to me: "I knew this wealthy funeral director who owned a number of homes. He was a raunchy kind of guy and when he got really drunk, he'd get pretty disgusting. He had a small circle of associates who worked for him and who enjoyed the same things he enjoyed. He'd throw these after-hours parties where bodies were laid out and they could do anything they wanted, as long as the corpse didn't suffer a lot of damage. Apparently they were going to be prepared for viewing the next day, so there wasn't time to cover anything up."

"Why did they let you see it?" I asked. "I would think they'd keep it a secret."

"I think they just didn't care. I had workers quit, I can tell you that, after seeing some of the things that went on. These parties were orgies. Everyone would get naked and run around using food and anything else they could find to be really outrageous. They'd pour bottles of wine down the throat of a corpse to watch its stomach bloat and then see if it would leak out anywhere. And if they had a young male or female, they'd pose the corpse over a couch and line up for a gang-bang."

"Why didn't anyone report them?"

"I think they were pretty careful to screen who got into these parties. There was some kind of initiation in the funeral home. First they'd say some things to see how a new guy reacted, and then they'd dare him to do things to a fresh corpse. If he seemed okay after a few times, he'd get told about the parties. It was quite a little club."

Such things are likely rare, but I did hear about another get-together that involved injecting corpses with fluorescent fluid. They'd set the corpse up against a wall, turn out the lights, and play black strobe lights onto it while they turned up the music.

Corpse abuse isn't limited to parties. There are handlers of the dead who have no regard for postmortem dignity.

In 1999, Ohio funeral home worker James Harber left the embalmed body of an eighty-eight-year-old man in a hearse while he made a visit to a topless bar. He was transporting the corpse from a funeral home that was 165 miles away to the one where he worked, so he decided to take a break at The Candy Store. He parked the minivan and walked away. A patrol officer spotted the uncovered body lying on a cot in the unlocked vehicle. He claimed that it was in plain view of anyone who might happen by. Harber was arrested on charges of corpse abuse.

Then there was the investigation of the Glebe morgue in Sydney, Australia, in March 2001. Allegedly, the staff had allowed physicians to use corpses for medical experiments. One former worker said he'd seen a pathologist stabbing a corpse to study blood spatter patterns, and another corpse was hit with a hammer to get results on blunt-force trauma. Sometimes organs were removed without obtaining permission and sent to research labs. The morgue boss admitted that approximately one thousand bodies left the morgue each year without a brain, which to his mind was just routine postmortem study. Apparently, the silence of relatives was taken as consent, and this discovery will likely inspire new laws.

Sometimes even a funeral home can be a Little Shop of Horrors. It was rumors that led to the downfall of the Lamb Funeral Home in Pasadena, California. In 1986 people complained about the way David Sconce was running his cremation business—the intense black smoke coming out of the stacks, the overwhelming stench, the impossible numbers of corpse disposals—and a full-scale investigation eventually led to twenty-one felony charges against him.

Highly ambitious, Sconce packed his prep room with bodies nearly to the ceiling. He'd pick them up daily from a large group of area funeral homes that hired him for cremation. As a side business, he harvested organs to sell to schools and labs. Generally he tricked relatives into signing the permission forms, and they often did not realize what they were allowing him to do.

Sconce threatened violence against competitor and employee alike. His "tissue technicians" quietly went about their business harvesting eyes and other organs as Sconce ripped gold fillings from the teeth of cadavers.

None of his clients realized just how many others he had, or they might have had second thoughts. Since he offered a price that couldn't be beat, no one asked questions. In short order he'd managed to orchestrate a cremation monopoly, and when he needed new facilities, he set them up illegally. He sent masses of bodies together into giant ovens disguised as ceramics kilns—as many as thirty at once. His employees were instructed to just keep shoving them in, as many as would fit. The commingled ashes were weighed and put into individual bags, and given back to people who were under the illusion that their loved ones had been cremated in a separate process.

When investigators finally came in to have a look, they found behind his "ceramics" building a sludge pile of human fat and cans full of remains. He shouldn't have been in that neighborhood, let alone leaving remains in the open.

Astonishingly, the judge eliminated half of the charges and allowed Sconce to serve a minimal amount of jail time. Even worse was the fact that his mother had masterminded the thefts and organ sales, but apparently believed that she'd done nothing wrong.

Unfortunately, theirs was not an isolated case. A funeral director in Arizona was arrested for taking valuables off corpses, and yet he's still in business. I'd also heard about the arrest of sixteen city morgue workers in Philadelphia for theft from the dead over a ten-year span. They got caught using some pur-loined credit cards. Apparently, when they went out for a re-moval, they saw this as an opportunity to explore someone's home and remove other things as well.

So it appears that there's some degree of truth in the stereo-type of the callous body handler. I heard about others along the way and already knew the story of John Wayne Gacy, one of the most notorious serial killers of the century. When he'd lived in Las Vegas as a young man, he'd worked as a janitor in a funeral home. There were rumors that he'd taken advantage of the pri-vacy at night and the fact that corpses can't fight back to fulfill his sexual needs. Later diagnosed as a psychopath with no sense of human attachment, he'd had no qualms about fondling the dead. He apparently also had no qualms about turning people into corpses, or of having them around him.

Years later, in the 1970s, he lured boys and young men to his home in Des Plaines, Illinois, tricked them into bondage de-vices, and then raped and strangled them. He admitted to keep-ing at least one in bed with him. Getting rid of the bodies was no problem. He just dumped them in the garden or the crawl space under his home, covered them with quicklime, and ex-plained away the noxious odors to his wife and friends as "mois-ture." When police finally investigated the disappearance of one young man, they searched Gacy's house and found seven bodies down below. More turned up in mounds in the garden and the floor of the garage. Four others he'd thrown into the river. In all, they convicted Gacy of thirty-three counts of murder, and he was finally executed.

A case of the fraudulent handling of a corpse turned up in

Florida recently when a funeral director murdered his wife. Since he buried bodies every day, he figured how hard would it be to dispose of an extra corpse? He and his wife of two years had been having problems and he just wanted to put it all behind him. He'd seen her walking with other men and he'd had enough. One night they had an argument that pushed things too far. She went to bed, and while she slept, he grabbed a steak knife from the kitchen drawer and stabbed her over and over until she was dead. He then took the body to his funeral home and stored it in a cooler. That was simple enough. Then, when he scheduled a closed-casket ceremony for an eighty-nine-year-old woman, he put his dead wife at the bottom of the casket, placed the mattress on top of her, and put the other woman's body inside. He closed the casket and buried them together. He might have gotten away with it, too, except that he ended up confessing to the police. The grave site was exhumed and, sure enough, the two bodies were found in the same casket.

Necrophilia

The movie *Kissed,* directed by Lynne Stopkewich, is about Sandra, a young woman who's erotically attracted to dead bodies, so she finds work at a funeral home. What she does there when she handles the dead is more passionate than anything she can find in life, even when she gets involved with a man. She thinks that by having sex with beautiful young deceased males, she can channel their spirits to a better place. "I'm consumed," she says.

The poet undertaker, Thomas Lynch, points out in *Bodies in Motion and at Rest* that both sex and death are "horizontal mysteries" that possess similarly disconcerting effects. Perhaps that's why some people so closely link sex and death, such as the W. W. Chambers Mortuary, which once issued a calendar featuring a nude embalmed female corpse under the legend "BEAUTIFUL BODIES BY CHAMBERS."

Some funeral directors assured me that now that video cameras are placed in many embalming rooms, necrophilia is rare to nonexistent, yet others said there are ways around anything. In fact, at the end of 2000 there was a lawsuit pending against an Atlanta-based funeral home about sexual high jinks with corpses in the receiving room.

"You won't get any stories from people like that," I was told. "They're pretty secretive."

On the contrary, I found several people who profess some form of necrophilia who were not only happy to talk but were quite generous with their tales. They know that people call them "sick" and "disgusting," but they also know where they find their most pleasure. And not all necrophiles work in funeral homes. In fact, most do not.

John Pirog lives in Michigan and has dedicated himself to promoting necrophilia. He keeps track of the most notorious necrophiles in history, such as Jeffrey Dahmer, Karen Greenlee, and Sergeant François Bertrand. He's also devised a list of "Necrophile Principles" and a somewhat disturbing "Guide to Tactful Funeral Home Visits."

I asked him why he had decided to engage in what he calls a "necrophilia outreach."

"No one else was doing it," he said, "and I felt that necrophiles need a rallying point. Over the years, I've met many folks who profess an interest in necrophilia. Unfortunately, most of us have only had limited experience with actual corpses. In Michigan, the regulations that cover mortuary science are rather strict, so finding work in any type of mortuary setting isn't that easy."

"Did you get attracted to corpses from some experience?" I asked.

"No, not really. I think I've always been a 'latent' necrophile, although it did not come to full fruition until around 1994." He dismissed the idea of the romantic necrophile, explaining that those who went all out in their sexual desire had a better grasp of the experience. "My own personal necrophilic feelings are more akin to those of the nineteenth-century French necrophile Sergeant François Bertrand. He mutilated and dismembered the bodies he exhumed."

I had a look at the profiles of the people he most revered and had to admit that these were pretty hardcore:

1. Sergeant François Bertrand liked to dissect animals as a child and had violent torture fantasies as he grew older. In 1849, he dug up fresh corpses with his bare hands from the grounds of Père-Lachaise and Montparnasse cemeteries in Paris in order to have sex with them. He'd also disembowel them and leave them strewn about the cemetery, and forensic evidence indicates that he chewed on a few. His youngest victim was only seven. Although he was caught and convicted, he served only one year in prison. He claimed that he couldn't help what he did; it was a compulsion.

2. Henri Blot was twenty-six when he began digging up graves in France. A ballerina had died and he pulled her from her grave to penetrate her. When he was finished, he fell asleep, waking only when the groundskeeper came upon him inside the grave. The corpse had obviously been ravished, so he was arrested. This was apparently his second such episode, and in court he reportedly said, "Every man to his own taste. Mine is for corpses."

3. Victor Ardisson was a mortician who reputedly had sex with over one hundred corpses in his care. He sometimes dug them up and took them home, and it was there that police found the decaying body of a three-year-old girl. He'd heard that she was ill and had fantasized endlessly about her corpse. When she died, he'd stolen her from a graveyard and had performed oral sex on her in the hope of reviving and restoring her. Then he kept her next to him when he slept. He also had possession of the head of a thirteen-year-old girl, which he referred to as "my bride," kissed from time to time, and kept on his bedside table.

Necrophilia is an erotic attraction to corpses. The most common motive cited by psychologists is the attempt to gain possession of an unresisting or nonrejecting partner. The activity fits the psychiatric diagnosis of "Paraphilia, Not Otherwise Specified," although I've met a few people who are, in Shelley's words, "half in love with easeful Death" and reject such a shallow approach to what they feel and do. Some merely fantasize about this; others take a more active role—even to the point of killing someone. Dennis Nilsen, an otherwise meek and gentle man who lived in London, murdered fifteen men and kept their decomposing corpses in his home—even his bed. He just did not want them to leave, he'd explained.

History offers several accounts of this activity, including the fear ancient Egyptians expressed that embalmers would violate their deceased wives. One legend states that King Herod killed his wife and then had sex with her for seven more years.

Supposedly, if one can judge such a secret activity, necrophiles are primarily male (about 90 percent), but one female apprentice embalmer claimed that during the first four months of her employment, she had sex with around forty corpses. She admitted that she could not achieve satisfaction with the living, in part because she had been molested once and later raped. She could express herself to corpses without fear.

Contrary to common belief, most necrophiles are heterosexual, although about half of the known necrophiles who kill are gay. Only about 60 percent of necrophiles have a diagnosed personality disorder; 10 percent of these are psychotic. The most common occupations through which necrophiles come across corpses include hospital orderly, morgue attendant, funeral-parlor assistant, cleric, cemetery employee, and soldier—although I must point out that the majority of people thus employed are *not* aroused by corpses. (One woman who was engaged to a coroner did tell me that before they made love he always asked her to take a cold bath.)

Most corpse violations occur prior to burial, but there have been cases where the corpse is disinterred from a cemetery plot. In 1985, a fifteen-year-old girl was buried in Italy after she died from a head injury. Two days later, her grave was discovered open. She was lying on top of her coffin, her white dress lifted up over her hips. An examination indicated that she had been anally penetrated, and the two shovels left at the grave site indicated that more than one person was involved.

◆ *Deviance*

One of my first up-close-and-personal encounters with necrophilia was with a young man who identified himself as a vampire because he loved the taste of blood. He went by the name "Anubis"—the Egyptian god of death rituals. He and I went together to an embalming room to look at the instruments so that he could explain something he'd witnessed as a mortuary assistant. He himself loved to work with corpses, but he thought that the head embalmer, "John," had taken things over the top.

John once had the experience, he told me, of actually meeting a young woman who was dying from leukemia and who would soon be in his care. Her name was Laurie, and she was twenty-four. She came in one day to make her own arrangements.

"John arranged her pre-need appointment," Anubis said, "and sealed the arrangements with the customary handshake and escort to the car. John patted her arm in mock consolation, a stall tactic designed to aid his memorization of her body movements."

Some time later, John invited Anubis to come and witness what he was about to do that evening, after hours.

"I arrived late the night Laurie's body came in. John directed my attention to the tabled remains. His hands caressed the body through the sheet, and then he yanked the sheet down, dramatically revealing the as yet unchiseled work of art."

What came next surprised him.

"John leaned down and pressed his lips against Laurie's cold flesh. His kiss was passionate and suggestive of some response. He treated her as though she were kissing him back. He spoke to her softly, nibbling her blue earlobes and smiling. He fondled her breasts and ran his hands across her belly, while his fingers slid in and out of her navel. He became increasingly aroused, removing his shoes and clothing. When fully naked, he leaped upon the table and straddled the corpse, with feet and hands ceremoniously placed on her thighs and shoulders. He was like a mosquito positioning itself for the insertion of its stinger, psychically linking himself with Laurie's spirit.

"He lowered himself into Laurie and began grinding his sweaty body against her icy frame. John seemed an unholy demon eagerly feasting upon the dead like a vulture picking at a carcass. His orgasm was violent, a literal blast from his loins into the cold, dry, and nonresponsive receptacle. He then dismounted from his icy lover and, kissing her sweetly, said goodbye."

As he told me this story, he watched to see my reaction. All I could think was that when my time comes, I hope I die in the woods someplace where no one can find me.

After that, I came across a rather famous, but no less hideous, tale than the one I'd just heard. It starts in 1931, when fifty-six-year-old Count Carl von Cosel, a radiologist, became obsessed with one of the tuberculosis patients in a hospital in Key West, Florida. Her name was Maria Elena de Hoyos, and she was a beautiful twenty-two-year-old woman. Von Cosel wanted her to marry him, but she died. Fearing contamination of her body from groundwater, he built a mausoleum for her in the cemetery and preserved her in formaldehyde. Unbeknownst to her family, he would sit and have "conversations" with her. He even left a phone next to her so he could speak to her while they were apart. When his routine nocturnal visits finally proved inadequate, he secretly moved her to his home.

Von Cosel brought in a regular supply of preservatives and perfumes, but she began to deteriorate. Using piano wire to string her bones together, he replaced her eyes with glass eyes and her rotting skin with a mixture of wax and silk. As her hair fell out, he used it to make a wig to put on her head. Stuffing her corpse with rags to keep her from collapsing and dressing her in a bridal gown, he kept her by his side in bed, even inserting a tube into her decrepit corpse to serve as a vagina for making love. He also played a small organ to her as she "slept." He did this for seven years, until de Hoyos's sister came upon her. Horrified, she called the police.

Von Cosel was arrested, but the statute of limitations had run out on his crime of grave robbing, so he was set free. He moved to central Florida, where he sold postcards of Elena. When he eventually died in 1952, he was found in a room with a large doll in his arms that was wearing Elena's death mask.

It isn't only males who indulge. I had heard of female embalmers handling bodies like this, and I was curious about how they managed it. I knew that a friend of mine, Mark Spivey, had worked in a funeral home once, so I asked him if he'd seen any clandestine sexual episodes. He said that he certainly had. One of the incidents he described was orchestrated by a woman named Debbie. They were friends, and she'd told him of her activities with corpses. Curious, he asked if he could watch sometime, and she said she didn't mind. So one night they got together in the prep room.

"She used to tell me," said Mark," that 'Beautiful men without souls are a dime a dozen, but beautiful men without both soul and breath are a treasure.' She had come to prefer beautiful but dead males. Now she had one, and she claimed that the one that had just been brought in was the most beautiful specimen she'd ever seen.

"She'd been saving the hydraulic pump for just such an occasion and now, she said, her victory was at hand. I watched as she inserted a thin plastic tube into his groin and sutured it in place. The scalpel slipped and I cringed. He couldn't feel it, but I almost did. Finishing the last suture, Debbie stepped back and squashed the pump pedal by her foot. The pump inflated the dead phallus, filling its tissues with hydraulic fluid. It got hard, very hard, and then she turned the clamp on the pump to maintain the pressure on the swollen organ.

" 'Just look at him,' she mused. 'Look at his beautiful face, those pouty lips, his empty eyes, his hard dick . . . just like every other man I've ever known.' She paused, then said, 'Only now, I'm in charge and you won't leave until *I'm* finished.' "

"Weren't you uncomfortable?" I asked Mark.

"I was. I realized that if I were the one lying there on the table, she'd be doing the same thing to me."

"So once she had him pumped up, what did she do next?"

"She undressed and draped her gown across the end of the table. She climbed on the table and positioned herself over her dead lover's cold erection, staring into his eyes. I guess that she was imagining his sweat-laden face, his heaving chest and racing pulse. Then she lowered herself onto his bluish-white cock and felt the icy pole push deep inside of her. She closed her eyes, moved into a rhythm, and moaned.

" 'All of them have to die,' she whispered, like she was in some kind of trance. 'Sooner or later, they all die.' Then she climaxed hard and long, raking her fingernails across the unresponsive skin on his chest. She slapped his face from side to side and twisted his nipples. When she was done, she climbed off him, and then bent over and swallowed him, tasting her own still-warm fluid."

I winced. "And then she prepped him?"

"No. Since she didn't need to have him ready for two more days, she saved the rest for later. Wheeling him into the storage room, with his penis still erect, she locked the freezer door. 'Just like a frozen dinner,' she said. 'Just pull it out and it's ready to go.' "

I let this episode sink in for a moment, and then I asked, "Was that the most dramatic incident you've seen, or does it get worse?"

Mark shook his head in a bemused way. He smiled and I could tell that something else was coming.

"I knew this guy once," he said. "Let's just call him John."

I laughed. Another corpse-loving "John."

"Anyway, John was in love with a younger guy, almost a boy, really. He was sixteen, but he was dying of cancer. Sometimes he mowed the lawn for John, and John told me that he tipped him really well because he did such a good job and because he always took off his shirt.

"John couldn't believe it when the boy's father asked him to do the funeral. The man wanted his son to be embalmed and he wanted John to do it."

Mark asked him how he felt about it, to which he responded, "Damn lucky. I could finally touch him."

Then John told Mark what he'd done.

" 'I washed the body first. I had washed hundreds of cadavers but none so carefully as this one. I caressed every inch of his bluish skin and gently lifted him on his side to wash his back, legs, and butt. I slid my finger inside his rectum . . . not something we do . . . but I wanted to feel my own flesh inside him just once. Something else we never do is wash the genitals. We usually just spray an antimicrobial mist and go on . . . but not this time. I washed his cock and balls and scrubbed them so hard I thought I could feel them filling with blood and getting warm.

" 'After I finished the preliminary procedures, I began to eviscerate his internal organs. Using the hydo-aspirator, I took great care in removing his innards. It upset me to be the one crushing and vacuuming out the soul of such a person, so I decided to stop when I came to the heart. It was then that I thought of a way to connect with my handsome yard boy.

" 'I used a small scalpel, the one I use to make the incision to raise the carotid for infusion, and cut into the left side of his chest and removed his heart. Placing the young and powerful organ on the table, I cut a small opening in the right ventricle. I was already excited just from working on the naked body, so it was no trouble to pull out my penis and jack off.

" 'Just when I was about to shoot, I stuck the head into the opening I made and emptied into his heart. I rammed my head against the backside of the heart wall, wanting to get some of his blood inside the little hole at the end of my dick. I wanted something of his inside of me. Then I realized I would just piss it out, so I cut a small piece out of his heart and swallowed it. That way I was sure to have him somewhere in me all the time. Sewing up the opening in the heart, I put it back into his hollow chest cavity and closed the hole.

" 'My seed would forever live in his corpse. His body will last forever—I made sure of that—and so will my sperm, dried and stuck to the inside of his perfectly pickled heart. This way, I'll always feel like I have his heart. I wouldn't have in real life, you know. It was the best job I've ever done on any body. His parents were so thankful, they even came over and brought me a gift. They said, "You took care of him as though he were your very own son." I smiled and said nothing, knowing that my sperm was in the flesh of his heart and that in that very physical way . . . he was now my son, too.' "

Mark ended his tale and we looked at each other. I was squirming inside. I had started on my journey with the idea that learning more about death and death care would make me feel less anxious. This kind of story brought it all back.

Finally I broke the silence. "How many bodies do you think have been violated like that?"

Mark shook his head. "We'll never know."

"It can't be that many. Can it?"

He just shrugged.

◆ *Necromancy and the Necromantic*

Having heard enough about sex and the embalming table, I decided to try to find a necromancer—a person who supposedly gets knowledge from communing with the dead. Necromantic rites, which generally take place in a cemetery or tomb, are a special means of divining or bringing forth energy by using the dead as a channel. Necromancers assume not only that the soul survives the body, but also that disembodied spirits have a superior knowledge or power that they can deliver to us if we approach them correctly.

Some form of necromancy was practiced in every nation of antiquity, particularly Rome, Persia, and Egypt. Some people believed they were calling forth malignant spirits, while others saw it as a form of white magic meant only for benevolent purposes. It was often the case that the person who summoned a soul simply wanted the mystery solved of how that person had died—particularly if he or she had been murdered. The rituals often involved animal sacrifice, but there were times when children were killed, too, in order to read their entrails. Necromancers who could make accurate predictions were held in high esteem, although when King Saul had the witch of Endor conjure up the spirit of Samuel, the dead prophet repaid him with an accurate foretelling of his imminent death. In later years, Christians were told that demons imitated the spirits of the dead and were not to be trusted for messages. Eventually necromancy was relegated to those forms of paganism that opened the door to the devil.

During ancient and medieval times, the same circles drawn on the ground that were used to conjure up demons or the devil himself served to draw forth the spirits of the dead. However, the necromantic ceremonies were performed in the cemeteries around a sarcophagus, and the dead person's name was used rather than the name of a demon. If he or she did not appear as summoned, the corpse was then exhumed for examination. Often the deceased were used as part of the festivities in some nocturnal feast.

Some practitioners used a "necromantic bell," which an eighteenth-century manuscript describes how to make (don't try this at home): The name THETRAGRAMMATON was inscribed on the bottom and above it were symbols for each of seven planets. Then ADONAI, and lastly, JESUS. Around the bell were carved the names of the various planetary spirits that would assist in the sorcery. The person who desired "confluence" with this bell—that is, to make it work for him—must cast into the correct shape an alloy of lead, gold, tin, iron, copper, mercury, and silver, and do it on the day and hour of his birth.

When not in use, the bell had to be kept wrapped in a piece of green taffeta, but once it was taken out, it was ready for the summoning. The way to use it was to place it in the center of a grave and leave it for seven days so it absorbed from the earth the character of the deceased. That made it effective as a channeling device when rung during the conjuring ceremony. The appropriate time for performing a ritual was within a year of someone's death, as that was thought to be how long the dead hovered around their bodies.

When I told people I was looking for a necromancer, I found the superstitions still very much alive. A few friends thought I was going to end up possessed. However, I finally discovered *The Necromantic Ritual Book,* by Leilah Wendell.

The basic idea is that some people need ritual to "connect with certain specific forms of energy"—in this case, death energy, or the "current of transition." *Necromantic* in this context is about the romance of death, as opposed to *necromancy,* which is more strictly about conjuring. Doing it right can enable one to share consciousness with Azrael, the Angel of Death—to experience his perspective—and Wendell warns that some of these practices are "not for the faint of heart."

She implores readers to consider what they are doing before trying out anything in the book. One must become comfortable in death's presence, not afraid, so it's a good idea to find some way to get close to death—apart from murder, that is.

I had been present at the deaths of several friends, been in a morgue, seen an autopsy, lingered in cemeteries, and had been immersed in numerous murder cases. I guessed I was ready, although I have to admit the idea of being alone in a mausoleum overnight (one of her suggestions) disturbed me. Still, I went right to the ritual of "High Necromancy," because I consider myself to be pretty tough about matters of darkness. At least, I thought so before I read it.

Right away I felt the serious nature of this enigmatic rite, called "a ritual of passion and devotion." It's reserved for the practitioner who seeks "intimate consciousness" with the Angel of Death. In other words, you must have a pure desire to be "Death's empath." If your motives are anything less, you'll likely be interrupted and expelled, but if you do it right, you'll get the "high" of your life.

The requirements include finding a very private cemetery, because in some states the practice about to be described is illegal. This cemetery must provide access to a corpse without the use of force or vandalism, such as in a mausoleum. Then you need to familiarize yourself with the comings and goings around this cemetery, particularly of the caretakers. (Some have guard dogs at night, so don't even think about it!) The ritual itself takes two days (or rather, a day and a night), beginning and ending at twilight. Wendell notes three specific dates when the

"West gate" is most likely to be aligned with our world. To make this work, you need to bring:

- a holly berry candle

- a jasmine incense cone

- a piece of amethyst

- an "athame" (blade)

- some milk mixed with a few drops of your blood

- a rock and a wedge to keep a mausoleum door tightly closed once you're inside

To get access to a corpse, you may need a casket key, which she says can be obtained from hardware stores (look for a hex key).

Now the ritual, which best involves a fairly decomposed corpse. This is used as a catalyst through which Azrael will emerge. You do various things with the corpse, such as:

- Light the candles.

- Pour some of the milk into its mouth.

- Take its hand.

- Extinguish all candles.

- Wait silently.

You may feel a chill, a sense of presence, dizziness, melancholy, or intense sensuality, any of which are signs that Azrael has arrived. Then you may be transported into a new experience. "When you return to the present world," Wendell promises, "you will understand those things hidden behind the scenes."

To prepare to depart:

- Savor this experience like a fine wine and allow the aroma to linger.

- Relight the candle.

- Kiss the catalyst.

- Thank Azrael.

- Make sure everything is restored to its previous state, although you leave the candle burning.

- Don't exit before the glow has left the amethyst.

- Use the athame to pierce the corpse's heart, leaving the blade behind. (This prevents both you and Azrael from being haunted by each other's dreams and also seals the gate.)

I read this ritual and thought, okay, maybe I wasn't ready. The idea of kissing a corpse was pretty daunting, not to mention that I've never been much of a magician.

However, I wanted to know if Leilah Wendell herself had actually done these things, so I made arrangements to meet her.

The Person Most in Love with Death

On Magazine Street in uptown New Orleans sits a purple-and-black house known to lovers of darkness as Westgate: House of Death. The name *Westgate,* symbolizes "the journey home." Each of the four gates is a transitional point. The east gate represents birth and the west gate the shedding of the body.

It's evident that Leilah is completely absorbed in her devotion to what she calls "death energy," because in two decades she has grown from the author of a couple of books into a publisher. She also owns an art gallery and museum entirely dedicated to death. In her autobiography, *Our Name Is Melancholy: The Complete Books of Azrael*, she claims that at the age of four, she encountered an entity that she believes was the Angel of Death—one of the seven archangels. Although the visitation terrified her, she eventually accepted it. This rather romantic entity let her know that she was his physical manifestation, and he continued to enter her life in various forms.

From him she learned how to send out her astral spirit, and then she had a series of supernatural experiences. More to the point, she moved ever more deeply into the aesthetics of death via a passionate lovers' tryst, and she now devotes herself to being a full-time prophet for Death's spiritual messages.

Apparently others share her experience, because she has received hundreds of letters from people detailing their own encounters with Death in personal form, such as this one from a woman in New Mexico:

> About a year ago, I attended the funeral of an acquaintance of my husband's. I had no choice but to bring my children (ages five and seven) with me to the graveside service. . . .We were standing around the grave site while people took turns eulogizing the deceased. All of a sudden my five-year-old started walking off as if he saw something standing at the foot of the grave. Everyone went silent as he walked up to a point, looked up and started talking, as if to a real person,

*but I saw no one. It was an awkward situation, as I couldn't
pull him away from this spot. A bit frustrated, I asked him,
"Who are you talking to, anyway?" His reply unnerved
everyone: "I'm talking to the really tall skinny man in the
long black coat." Of course, no one else saw this, or at least
no one else admitted to seeing this. After the service I asked
my son what they talked about. His reply was, "I don't
remember, but he talked without moving his lips."*

I parked in front of Westgate and walked up the steps just as a
young man with long, wavy brown hair opened the door. I figured him to be Daniel Kemp, the curator and the author of *The
Book of Night*. He shared with Leilah the desire to be close to
Azrael. Daniel silently beckoned me in to a dark hallway and I
looked up the wooden staircase to the right to see a slender,
pale woman with brown hair and wire-rimmed glasses seated
on the steps. She was dressed in black.

I knew at once that it was she. This was Leilah Wendell.

Before we talked, she invited me to look around her museum
of death. She had spent most of her life researching and depict-
ing personifications of death, and that was evident in what I saw.
However, the museum is not as morbid as it may sound (or
maybe my tastes are more morbid than I realize), because I
found much in the room to be enchanting, even inspiring. On
the black walls were detailed acrylic paintings of various scenes,
from shrouded figures moving into the woods to a skeleton em-
bracing a beautiful woman. Right in the middle of the floor was
a huge statue of a winged golem. In front of this creature, Leilah
told me, a businessman had once sat silently for hours until he
began to cry. When he left, he simply said, "Thank you."

In another dark room, I found a case full of silver jewelry and
graveyard replicas, such as the famous skeleton image from the

Susanna Jayne tombstone in Marblehead. There were no kitschy Halloween effects, no fake spiderwebs or children's props. It seemed a place where people who loved death culture were taken seriously.

In a back room was a tall, life-size papier-mâché sculpture of the Angel of Death, Azrael, holding a swooning woman in his arms who looked very much like Leilah, and close by were more paintings and a shelf full of books about death. Leilah told me that a former army colonel had come here to ask a rather unusual favor: He was dying from cancer and wanted to stay in the gallery until his time came so that he could be in Azrael's presence. Leilah granted his wish, and four days later that's just where he died.

Leilah invited me into her room, painted black, so we could talk. Her black cat followed us in. I sat on her black-covered couch, while she sat on a bed.

She quickly told me that necrophilia has too long been vilified and that she wanted me to understand what it means to love Death. It's a choice, a preference, not a compulsion. Love should not involve violation, but it does involve getting close to death. In other words, this was necromancy, as in "romantic" or necro-eroticism, not necrophilia as in the use of a corpse for selfish sexual gratification. The idea was to achieve emotional and spiritual intimacy with death.

"I don't like life energy," Leilah stated. "I avoid it." In answer to the next obvious question, she admitted to several attempts to finally join Death, but then realized that she had a mission on Death's behalf. It was given to her to spread the word.

I noticed a sculpture hanging on the wall of a figure wearing a death mask. Leilah was happy to explain it.

She said that she'd once been a funeral director and a morgue assistant, and her primary interest had been in disinterring bodies for reburial—a passion she shared with a friend, John (another John!). Together they were called "the Resurrectionists."

Leilah explained that she prefers the "moldy oldies" to the freshly dead and bacteria-laden "gooey Louies," so she got the surprise of her life when John pulled up to her home one evening in his customized "Deathmobile" (a cross between a hearse and long-bed pickup truck). He hinted that he had brought her "something special" for her twenty-eighth birthday, and he then pulled out a large package wrapped in a red bow. Carrying it in and placing it on her carpet inside, he urged her to unwrap it. She did, and before her lay a dessicated corpse, remarkably preserved. It had been exhumed from a pauper's cemetery and its disposition was pending.

"It's yours until Tuesday," he said. Then he left and she was alone with what had once been a man. The first thing she did was make a death mask, which became the centerpiece of the sculpture that I had noticed. The other thing she did—take the corpse to bed—I didn't quite understand until she had me read a piece she had written about one of her first erotic experiences with a corpse. It was one of her surprise encounters with Azrael.

This had occurred in a cemetery in New York, in a small mausoleum. She had come across an old tomb, so she entered it. To her right were three coffins and to her left a sarcophagus that held a casket. What happened next is best told in her own words:

"I reached down into the ravaged casket and began to peel the sheet from the dead form. Gently unwrapping the brittle cloth from the decaying body, I had to reach underneath the corpse to untie the twine that held the withered arms folded across the chest. As I bent over him, one of the skeletal hands slowly reached up and caressed my body, and then another clasped me in a cold embrace!

" 'Here as well,' I whispered to myself, as he pulled me down into the coffin until my lips pressed against his cool decay! The taste of his ancient love was like the bitter harvest of a dying bloom, and as sweet as an exotic nectar. I could feel the outline of his bones pressing up against my body. As I pulled the sheet away from his face I saw how magnificent and beautiful he was.

" 'Let me sleep beside you,' I whispered as I gently parted the veil of cobwebs that covered him and climbed into the open casket and lay down beside him. Drawing the fragile corpse into my embrace, I could feel my heartbeat pounding against the hollow chest. For the first time in a long time, I felt close again to my beloved Death. Somehow, the dead were able to serve as direct catalysts between Death and myself. Being near 'them' somehow brought me closer to 'Him,' and made me more fully love all aspects of what He is.

"Lying beside Him, I suddenly began to feel more intentful movement! The attempting of a definite embrace sent shivers through my soul. The arms that pulled me down were tightening around me!

" 'Come through, come through, my love.' I appealed silently to His attentive spirit. 'I call to you, my most beloved,' I said as I surrendered to His approaching gloom. The skeletal embrace became even more distinct, as the stirrings in my soul intensified. . . . The figure in my embrace was no longer a human remnant. The dust and decay had somehow reformed into something way beyond explanation! Something too incredible to describe. I knew it was my beloved Azrael. I could 'feel' His soul. By some magical means, He was able to alter the molecular structure of the decomposing elements and form them into a tangible chrysalis. A truly personified embodiment of Himself. A GENUINE, ANIMATED CATALYST!"

There was much more, but this was enough for me to understand that Leilah had certainly tried out the rituals that she had written in her necromantic spell book. She confirmed this. Being that close to a corpse not only did not disturb her, it put her into a swoon.

I felt the way I always do when someone tries to explain to me the ecstatic joys of financial planning—I was polite but I didn't really get it. Yet I couldn't help but be impressed with how open and without guile Leilah was. She freely told me anything I wanted to know about these activities, and yet she'd had her share of people denouncing her. She knew nothing about me, really, yet here she was just opening up. I have to admit, I found her to be a complex and intelligent person. Although I wouldn't want to climb into bed with a corpse, her view of necrophilia was certainly less brutal than what I'd heard before.

◆ Ghouls

Body snatching has long served the purposes of religious cults that need certain human organs for their sacred rituals. In 1604, King Henry made it a felony in England to steal a corpse for the practice of witchcraft. It still goes on in New Orleans and other places around the country where cults practice dark rites, but grave robbers nowadays generally aim to raid tombs for jewelry, stained glass, or funereal statuary. One person in Cincinnati, Ohio, even "stole" images. He convinced morgue workers to allow him to take photographs of corpses posed with objects like sheet music and syringes. Into the hand of a young girl's corpse he placed *Alice in Wonderland*, and he also got a shot of an infant autopsy before officials stepped in to stop him.

Despite the morbid nature of the activity, cemetery theft has a fascinating history that still reverberates. That someone has stolen artifacts from a mummy's tomb thrills us with the idea that the thief is now cursed and will meet some dreadful end. We have some feeling that corpses are aware of violations and possess the power to avenge them.

Grave robbing reached its morbid height during the eighteenth and nineteenth centuries, when it served the medical establishments. Physicians needed to dissect bodies to improve their knowledge, and it was better for medical students to work on their own cadavers than merely watch the teacher.

At first, there were no real laws, except prohibitions against taking a corpse's possessions, so grave robbers took care to leave behind the corpse's clothing. However, that situation eventually changed. With the understanding that doctors could better help patients with a proper grasp of anatomy, the government nevertheless balked at allowing them to just go dig someone up. Fines were imposed and grave robbers were arrested.

In Britain, legal restrictions were especially strict. Although King Henry VIII allowed four corpses to be taken annually by doctors from the gallows, the supply was terribly inadequate for the increasing number of students. In some places, students were actually required to supply their own corpses, and there was no place to get them but the grave. That pitted them against the authorities, which gave rise to secret dissection rooms and midnight requisition crews, otherwise known as the "sack 'em up" men. At first the students (and professors) went out to dig up the bodies themselves, but those with money hired "resurrection" specialists, who were quite adept at spiriting a body out of a grave.

Corpse theft grew into big business and professionals tipped sextons or gravediggers to let them know when a body would be available. They also followed funeral processions so they would know where to go within the next few nights. Corpses were best if removed within ten days of death.

The snatchers would set up a small tent to shield the lantern, use wooden spades or picks, and dig down only at the head end. Then they'd break open the top part of the lid to lift the body out with a hook and rope. Once it was laid out at the graveside, they tied it into a compact position, generally doubled up, and shoved it into a sack. Real professionals working in teams could dig down, get the corpse, and replace the dirt and sod in about an hour. One gang of ghouls had stolen nearly eight hundred bodies in two years.

Bodies were also taken for their valuables, their teeth, and their fat, which was used for candles. A few people even had the thought that they could take a famous body, such as that of Abraham Lincoln, and hold it for ransom. In 1876, a gang broke into Lincoln's tomb and managed to lift the former president's coffin out of the sarcophagus before Pinkerton detectives interrupted them. The coffin was then embedded in steel and concrete.

People caught on to the danger to their deceased relatives and took pains to protect them . . . and themselves! Many feared the anatomist's knife and did not want to end up on the dissecting table. Some turned to cremation as the only recourse, but there were other clever ways to outwit a ghoul. Since the snatchers generally put a hook through the jaw or a rope under the head and took a body out by the head and shoulders through broken wood, steel caskets began to be manufactured and sold. Some inventors also offered steel casket cages to place over the coffin, but since metal stayed around longer than wood, the cemetery managers frowned on these devices. They needed to recycle the graves. At any rate, the thieves figured out how to break through metal and get the corpses.

People also stood vigil until they felt sure that the corpse had decomposed sufficiently to no longer be of interest, but this was a wearing activity that required a number of people. One surgeon, who knew all too well what could become of him because he did it to others, insisted that he be wrapped in flannel blankets when he died and left in his rooms until he had decomposed. Only then was he to be transported to a cemetery.

A few cemeteries offered the service of a locked holding house, where bodies were laid out to rot past the point where medical schools would want them, but this was an extra service that cost money, so not everyone could afford it. Some undertakers sold harnesses to strap the body in place, while a few grave sites were rigged with bombs, guns, and gunpowder— and there were certainly fatalities. A medical student in Scotland tripped a spring-loaded gun and was instantly killed.

Once the body was delivered, a certain amount of haggling over price went on. Bodies were sold by the inch, and had to be in good condition—preferably without any marks on them. Sometimes only parts were brought in, or fetuses, and there was always the possibility that a fresh corpse harbored a contagious disease. Since the thieves were risking themselves, they expected to be paid well. If doctors and students wanted a steady supply from a reliable supplier, they had to negotiate well, because there was always another school looking for the same services. Resurrectionists were businessmen, after all.

Some did tire of digging up graves and figured that there was an easier ways of getting cadavers: murder. The term for a certain type of smothering that left no mark came to be known as "burking," named after William Burke. He ran a boardinghouse in Edinburgh, Scotland, with his partner, William Hare. Together they would get their victims drunk and then either grab them from behind in an arm lock around the throat or sit on their chests while holding their nose and mouth closed. In nine months, they managed to kill sixteen people and then sold them one after another for an average of ten pounds apiece. When they were caught in 1828, Hare agreed to testify against Burke. He went free while his wife and Burke's mistress fled the country. Burke was hanged, and in an ironic twist of fate, his corpse was turned over to the anatomists at Edinburgh University to be dissected. Thirty thousand people saw his execution and then his anatomized body, which was put on public display to deter others from mimicking his foul deeds.

The physician who was known to be Burke's regular customer went on unhindered with his surgical career.

Only two years after Burke was hanged, another team in London turned to murder as well. They'd already supplied well over five hundred corpses from cemeteries, and then they killed three people by drugging them and then lowering them into a well, headfirst. They, too, were hanged and dissected.

Women got into the act, too, when a pair of prostitutes killed a sick child. It seems that the child just wouldn't die as expected, so they played God. They were hanged for their deed, but the activity of selling corpses was so lucrative during certain times that one mother even tried to sell her own children.

The same conditions applied in the United States Anatomists were forced to pay someone to go dig up fresh corpses for study, and as recently as 1989, the discarded remains of some of these bodies from the nineteenth century were found in the cellar of a medical school in Atlanta, Georgia. As much as possible, the snatchers raided paupers' graves.

Body snatching still exists today in some places. Charlie Chaplin, the famous British comedy actor, died on Christmas day in 1977 at the age of eighty-eight. He was buried in the cemetery at Corsier-sur-Vevey, near Switzerland's Lake Geneva. By March of the following year, someone had stolen his corpse, coffin and all. His widow declined to pay a ransom of four hundred thousand pounds, claiming that he was with her in spirit and she didn't much care where his remains were. Over two months later, some ten miles from the cemetery, the coffin was found in a cornfield. Two men were arrested and the body reinterred in a concrete vault.

As late as 1992 in Columbia, there was an attempt to murder a man to sell him as a medical specimen. He alerted police, who came upon a cache of bodies and parts. A security person was arrested and convicted of murdering fifty people to sell them to the medical school for two hundred dollars apiece.

This past year in Nepal, a woman was caught selling skulls from graves as curio items, while reward money inspired body snatchers in Vietnam to raid dozens of graves to pass off the bones as the remains of American soldiers.

The theft of a body in Italy in March 2001 made international headlines. Legendary banker Enrico Cuccia, the father of Italian capitalism, had died the year before. A team of men broke into his tomb in northern Italy, lifted the heavy slab, and removed the coffin. While authorities pondered the motive, they received a message from a man who claimed that he'd taken the corpse because he'd lost a fortune on the stock market. When the economy improved, he'd return it. Not long afterward came a ransom demand of $3.5 million, and soon Giampaolo Pesce was arrested and charged. The corpse was located in its coffin in a hayloft near Turin, and restored to its tomb.

Thus, grave robbing still happens, but these days the term "body snatcher" applies to mostly unscrupulous undertakers who want to claim the state's burial allowance for the homeless or destitute. They "snatch" the bodies from morgues or hospitals, usually by some sort of bribery or manipulation that gives them an edge on this business. Since their primary motive is money, they're viewed as ghouls.

The Ultimate Fiend

Of all the stories I've heard about grave robbing and body parts, Ed Gein is the epitome of the ghoulish stereotype. Best known as the man who inspired Alfred Hitchcock's *Psycho* and Thomas Harris's Buffalo Bill in *The Silence of the Lambs*, his story unfolded in the 1950s as one of shocking proportions.

When police investigated him one afternoon on suspicion of robbery, they snooped around his farm and then entered his barn. There they found evidence that he'd been hunting, but not for deer. Hung from the ceiling, feet first, was the headless corpse of a woman, slit from her genitals to her neck, washed clean, with her legs splayed wide apart. It was quite a shock, but that was only the beginning.

Gein had been raised by a domineering mother on a farm outside Plainfield, Wisconsin. By 1945, his father and brother had died and his mother had suffered a stroke that left her paralyzed. She eventually died. Alone and socially inept, Gein devoured books on human anatomy and Nazi experiments. Then one day he spotted a newspaper report about a woman who had just been buried. He decided to go out and dig her up so he could have a look at a real body—a female one. He got a friend named Gus, who was a gravedigger, to help him open up the grave, which was near where Gein's own mother was buried. He continued with this habit, usually under a full moon, for the next ten years. Sometimes he took the entire corpse and sometimes just certain parts. He later claimed that he had dug up nine separate graves in three different cemeteries. (Police did not believe him until they went out to exhume the bodies . . . or tried to. Some weren't there.)

Storing the organs in the refrigerator, he made things out of the bones and skin of these women. Sometimes he had sexual contact with them (though he denied that), and eventually he just went ahead and dug up his own mother. What he wanted to do was become a woman—that is, to become *her*. Rather than get a sex-change operation, he simply made himself a female body suit and a mask out of the skin taken from his grave thefts, and he would wear this and dance around. He also sometimes wore it to dig up graves.

Finally he decided to get bodies that were more pliable. That meant killing someone. In 1954 and 1957, Gein shot two older women who resembled his mother and brought them to his farm. No one suspected a thing, but eventually the police decided to have a look. When they invaded his home, they found a horror house that contained four noses, several bone fragments, nine death masks, a heart in a pan on the stove, a bowl made from a skull, ten female heads with the tops sawn off, human skin covering several chair seats, pieces of salted genitalia in a box, skulls on his bedposts, organs in the refrigerator, a pair of lips on a string, and much more. It was estimated that he had mutilated some fifteen women and kept their remains around him.

At his trial, he was found to be insane, and he died at the age of seventy-eight in a psychiatric institution. Since he had never actually hunted for deer, neighbors wondered what had actually been in the packages of fresh venison that he'd so generously brought them.

Gein was buried in his family's plot, which is one of the cemeteries from which he stole bodies. On his gravestone it simply says, EDWARD GEIN, 1906–1984. A cross and two carved palm leaves decorate the front, and many tourists have come to chip off grim souvenirs.

Now someone has decided that what goes around comes around, and it's Gein's grave that has been violated. On a Saturday night in June 2000, some rather bold (and strong) thieves stole the two-hundred-pound gravestone. There were no leads, but police officers speculated that it would show up on some Internet auction site.

Afterthought

As I look back over these stories, I'm amused, disturbed, and delighted by the range of human behavior surrounding the subject of death. I'm grateful that we have people willing to help us with the rituals of life's termination, but saddened by those who feel that their own self-enrichment or gratification is worth more than another person's dignity. As I walk through a cemetery now, I understand the differences among the types of stones and monuments, and I can appreciate the many stories there are yet to be recorded. Although I traveled far and wide and talked to many people, I feel as if I've only scratched the surface. However, I know so much more about death now, from the serene aspects to the sinister, and about how people hope to be treated and remembered.

The obvious question now is, what shall be my own final disposition? Will I do pre-need? Will I decide to be buried? Will I have an elaborate epitaph?

In thinking it through, I find that I agree with the final words of Ethan Allan, who died in 1789. Upon being told that the angels in heaven were waiting for him, he said, "Waiting, are they? Well, let 'em wait!"

Resources

Ames, Lynn, "Burial Society's Sacred Task of Performing a Ritual for the Dead." *New York Times*, March 9, 1997.

Association for Gravestone Studies. "Gravestone Rubbings for Beginners."

Bahn, Paul G. "The Face of Mozart." *Archaeology*, March/April 1991.

Bouchard, Betty. *Our Silent Neighbors: A Study of Gravestones in the Old Salem Area*. Salem, MA: T. B. S. Enterprises, 1991; second edition, 2000.

Braidhill, Kathy. *Chop Shop*. New York: Pinnacle, 1993.

Carlson, Lisa. "Bones, Bugs, & Batesville." FAMSA newsletter, Spring 1999.

————.*Caring for the Dead: Your Final Act of Love*. Hinesburg, VT: Upper Access, 1998.

Coco, Gregory. *Wasted Valor*, Gettysburg, PA: Thomas Publications, 1990.

"Coroner's Office Workers, Others Charged with Stealing from Dead." CourtTV.com, Dec. 21, 2000.

Crary, David. "Hillside Cemetery Binds Halifax to *Titanic* History." *Detroit News*, March 16, 1998.

Crumm, David. "How to End Up Like a Pharaoh." *Detroit Free Press*, Aug. 5, 2000.

Culbertson, Judy, and Tom Randall. *Permanent Italians*. New York: Walker & Company, 1996.

————. *Permanent Parisians*. Chelsea, VT: Chelsea Green, 1986.

"Disaster Response Team." Discovery Channel production, September 2000.

Edgar Allan Poe Society of Baltimore. "Poe's Grave." www.eapoe.org.

Eggan, Dan. "Death Finds Life on the Web." *Washington Post*, May 17, 2000.

Florence, Robert. *New Orleans Cemeteries*. New Orleans, LA: Batture Press, 1997.

Geberth, Vernon J. *Practical Homicide Investigation*, 3rd ed. Boca Raton, FL: CRC Press, 1996.

Givry, Emile G. *Illustrated Anthology of Sorcery, Magic and Alchemy*. NY: Causeway Books, 1973.

Habenstein, Robert, and William M. Lamers. *The History of American Funeral Directing*. Milwaukee, WI: Bulfin Printers, 1962.

Hansen, Joyce, and Gary McGowan. *Breaking Ground, Breaking Silence: The Story of New York's African Burial Ground*. New York: Henry Holt, 1998.

Harrison, Ben. *Undying Love: The True Story of a Passion That Defied Death*. New York: New Horizon Press, 1996.

Hatch, Robert T. *What Happens When You Die? From the Last Breath to the First Spadeful*. New York: Birch Lane Press, 1995.

Horn, Miriam. "The Deathcare Business." *U.S. News and World Report*, March 23, 1998.

Hucke, Matthew. "The Chase Vault." www.ghosts.org.

Iverson, Kenneth. *Death to Dust*. Tucson, AZ: Galen Press, 1994.

Johnson, Steve. "WOW: On the Search of Graves." *The Cemetery*

Column, interment.net, March 15, 2000.

Jones, Constance. *R.I.P: The Complete Book of Death and Dying*. NewYork: HarperCollins, 1997.

Kinkopf,Vlad. "Nellie's Grave," in *Voices from theVault*, Devemdra P.Varma, ed.Toronto: Key Porter Books, 1987.

Knight, David C. *The Moving Coffins*. Englewood Cliffs, NJ: Prentice-Hall, 1983.

Lanci-Altomare, Michele. *Goodbye My Friend: Pet Cemeteries, Memorials, and Other Ways to Remember*. Irvine, CA: Bowtie Press, 2000.

Lynch,Thomas. *The Undertaking*. NewYork: Penguin, 1997.

————. *Bodies in Motion and at Rest*. New York: W.W. Norton, 2000.

Manhein, Mary H. *The Bone Lady*. NewYork: Penguin, 1999.

Massie, Elizabeth. *Sineater*. NewYork: Leisure Books, 1998.

Meyer, Richard E., ed. *Cemeteries and Gravemarkers:Voices of American Culture*. Salt Lake City, UT: Utah State University Press, 1992.

Michaud, Stephen, with Roy Hazelwood. *The Evil That Men Do*. NewYork: St. Martin's Press, 1998.

Mitford, Jessica. *The American Way of Death, Revisited*. New York: Vintage, 1998.

Murphy, Edwin. *After the Funeral: Posthumous Adventures of Famous Corpses*. NewYork: Barnes & Noble, 1995.

Parker, Suzi. "GetYour Laws off My Coffin." salon.com, Jan. 12, 2001.

Ramsland, Katherine. *Ghost: The Art of Shadowing Spirits*. New York: St. Martin's Press, 2001.

————. *Piercing the Darkness: Undercover withVampires in America Today*. NewYork: HarperCollins, 1998.

Russell, Davy. "Phantom Dogs." *Paranormal Bohemian*, Nov. 15, 1999.

Sloane, David Charles. *The Last Great Necessity: Cemeteries in Amer-*

ican History. Baltimore, MD: The John Hopkins University Press, 1991.

Stepzinski, Teresa. "Historians Seek Preservation of Slave Cemeteries." *Augusta Chronicle*, Sept. 7, 2000.

Taylor, Susan. "PluggedIn: Grave News Abounds on the Internet." AOL News, Dec. 26, 2000.

Wangemann, Garth W. "The *Titanic* Graves of Halifax." www.execpc.com/~reva/html3b.htm, Oct. 12, 1998.

Wasik, John. "Avoiding the Cemetery Trap." *Consumers Digest*, May/June 2000.

Wilkins, Robert. *The Bedside Book of Death*. New York: Citadel, 1990.

Wilson, Keith. *Cause of Death: A Writer's Guide to Death, Murder, and Forensic Medicine*. Cincinnati, OH: Writer's Digest Books, 1992.

Young, Gregory W. *The High Cost of Dying*. New York: Prometheus, 1994.

Zbarsky, Ilya, and Samuel Hutchinson. *Lenin's Embalmers*. London: Harvill Press, 1997.

Selected Web Sites

Tomb with a View *http://members.aol.com/Tombview/twav.html*
Tombstone Traveler *http://www.flash.net/~leimer/*
Adipocere *http://adipocere.homestead.com*
NecroErotica
   http://home.earthlink.net/~john30/public.html/index.htm
City of the Silent *http://www.alsirat.com/silence/index.html*
FAMSA-FCA *http://www.funerals.org*
http://www.findagrave.com
http://www.funeralguy.com
http://www.lifefiles.com
http://www.nfda.com
http://www.paylesscaskets.com
http://legacy.com
http://www.ashesonthesea.com
http://www.casketsonline.com
http://www.funeralstodiefor.com
http://www.pyramiddevelopment.com
http://www.FinalThoughts.com
http://www.gravestonestudies.org
http://www.westgatenecromantic.com

http://www.a-grave-affair.com
http://www.celebritymorgue.com
http://www.frugalfuneral.com
http://www.cemeteryshop.com
http://www.funeral-cast.com
http://www.cremationassociation.org
http://www.americawills.com
http://www.funeralplan.com
http://www.gravenews.com
http://www.buycaskets.com
http://www.TheCemetery.net
http://www.forevernetwork.com